Merrow Technical Library

Textile Technology

FLUORESCENT WHITENING AGENTS

Merrow Technical Library

General Editor: J. Gordon Cook, B.SC., PH.D., F.R.I.C.

Textile Technology

MTL/TT/1	Handbook of Textile Fibres. J. Gordon Cook. 4th Edition.
MTL/TT/2	Handbook of Polyolefin Fibres. J. Gordon Cook.
MTL/TT/3	Spinning in the '70s. P. R. Lord (Ed.)
MTL/TT/4	An Introduction to Warp Knitting. D. G. B. Thomas.
MTL/TT/5	Fluorescent Whitening Agents. A. K. Sarkar.

FLUORESCENT WHITENING AGENTS

A. K. Sarkar, B.Sc., B.Sc.(Tech.), Ph.D., F.R.I.C.

Cyanamid N.V., Rotterdam, Holland
(formerly of Hickson and Welch Ltd., Castleford, U.K.)

MERROW

Merrow Publishing Co. Ltd.

I.S.A. Building
Dale Road Industrial Estate
Shildon, Co. Durham
England

ISBN 0 900 54120 2 Clothbound
ISBN 0 900 54121 0 Paperback

Printed in Great Britain at the Pitman Press, Bath

CONTENTS

1

INTRODUCTION AND HISTORY

A quarter of a century ago fluorescent whitening agents were virtually unknown, yet today they are used extensively by industrial and domestic consumers alike. Most detergents—laundry or household—contain fluorescent whitening agents, and it is practically impossible to obtain white textile fabric or paper which has not been treated by one or other of them.

Nomenclature

Fluorescent whitening agents are known under various names, e.g. optical whitening agents, optical bleaching agents or optical bleaches, fluorescent bleaching agents, whiteners, brighteners, etc. In the Colour Index they are listed as fluorescent brightening agents. The ASTM[(1)] have suggested the term "Fluorescent Whitening Agents" for them, and for the sake of convenience and simplicity it is proposed to refer to them in this book as FWAs.

Early Work

Over a century ago, Stokes studied the phenomenon of fluorescence on a scientific basis[(2)]. In 1921, A. V. Lagorio[(3)] observed that the white content of colours can be enhanced by fluorescent dyestuffs. In 1929 Krais[(4)], using Lagorio's principle, showed that yellowed viscose was whitened when dipped in a solution of aesculin (Fig. 1.1) the glucoside of 6:7 dihydroxy coumarin, which is extracted from the bark of horse-chestnuts; the treated fabric was comparable with bleached material. This demonstrated that a substance yielding a colourless solution with blue fluorescence could compensate the yellow of the original material.

Krais's discovery could not be put to practical use, as aesculin had no affinity for cellulosic fibre; the white effect was lost when the substrate came in contact with water. It demonstrated for the first time, however, that a compound giving a colourless fluorescent solution having substantivity to fibres could be used in practice to enhance their whiteness.

aesculin

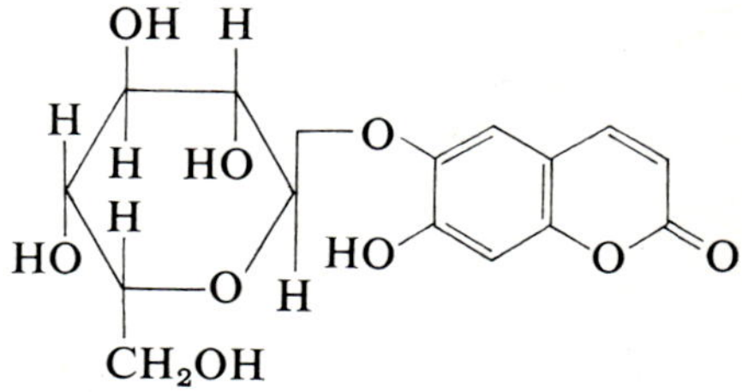

Fig. 1.1

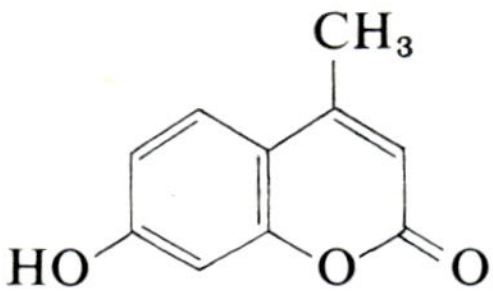

β-methyl umbelliferone

Fig. 1.2

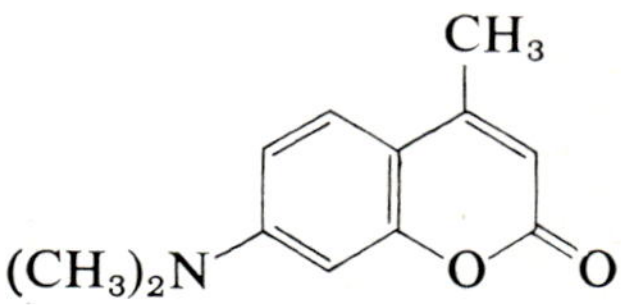

4-methyl-7-dimethylamino-coumarin

Fig. 1.3

Although it was apparent that coumarin derivatives of the aesculin type were of no practical value for whitening cellulosic textiles, research continued on the fluorescent derivatives of 7-hydroxy and 7-amino coumarins. In 1935, Ultrazell GmbH(5) and Hoffmanns Stärkefabriken(6) patented the use of β-methylumbelliferone (Fig. 1.2) and 4-methyl-7-dimethylamino coumarin (Fig. 1.3) under alkaline conditions for the whitening of cellulosic textiles. The effect was not fast to washing. The products were also recommended for the protection of food-wrapping materials from UV light.

In 1934, ICI[7] patented the use of diacyl derivatives of 4:4′-diamino-stilbene-2:2′-disulphonic acid (an intermediate used in making direct dyestuffs) for producing substantive fluorescent effects on cellulosic textiles and paper, and as UV screening agents for cellophane wrappings. In 1937, I. G. Farben[8] reported using bis-triazinylamino compounds, derived from the condensation of water-soluble mono- and diamino-compounds with cyanuric chloride, as light protecting agents by virtue of their strong UV-absorbing properties. Some of these compounds had properties similar to those of Krais's whitening agent, and one was marketed as "Ultrasan" in 1938.

In 1940, Bruno Wendt and coworkers[9] of I. G. Farben found that triazinylaminostilbenes of the following formula (Fig. 1.4) could be used from detergents and alkaline baths as substantive FWAs for cellulosic textiles:

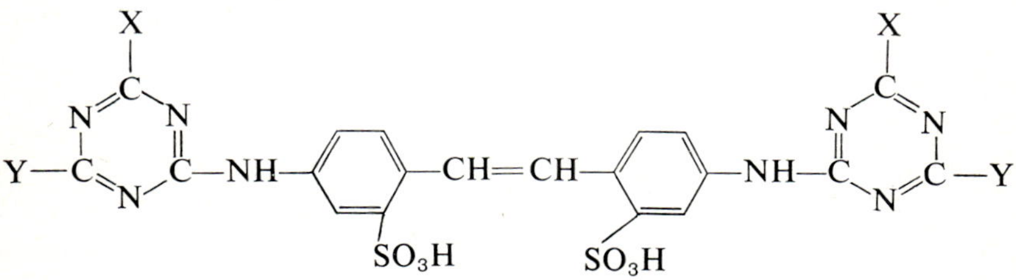

Fig. 1.4

Some of the compounds mentioned in this patent are sold in substantial quantities throughout the world today.

Since that time, rapid developments have taken place in this field. The earlier products were for use almost entirely on cellulosic textiles, but FWAs are now used to whiten a wide variety of products including all types of synthetic fibres and fabric. The development of FWAs presents a challenge to many disciplines. To the organic chemist, FWAs are complex compounds that present problems of synthesis; to the textile chemist they are white dyes which carry the added complication of dependence on the UV-content of the light source; to the manufacturer, they represent an expanding business worth (1970) some $100 million per annum. In the U.S.A. alone, annual production of FWAs in 1968 was estimated at 31,297,000 lb; sales were valued at $52.67 million(10). Most of the major chemical and dyestuff manufacturers of the world are engaged in the development and marketing of FWAs. By 1970, more than 2000 patents had been taken out, covering a wide range of aromatic and heterocyclic FWA systems, and claiming a wide variety of uses.

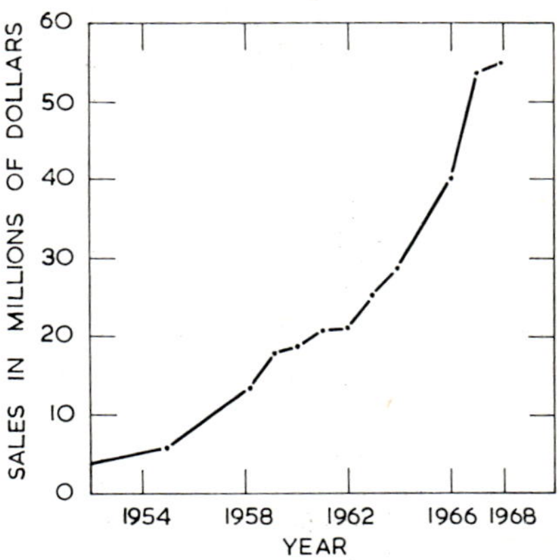

Fig. 1.5. U.S. sales of FWAs

Optical Bleaches

Bleaching is concerned with upgrading the whiteness of textiles, and FWAs have been described as optical bleaches.

Bleaching may consist of the removal of coloured impurities, their conversion into colourless substances or the physical compensation of colour. In chemical bleaching, impurities are converted by oxidizing or reducing agents into colourless products. Physical bleaching consists of the introduction of a complementary colour; the undesired hue thus being made optically invisible; it may be carried out either by

(*a*) the age-old "blueing operation", or
(*b*) using a fluorescent whitening agent.

Perfectly white bodies reflect and ideal black bodies absorb all light that falls on them. These ideals are seldom achieved in practice, most bodies reflecting some of the light that falls on them, and absorbing the rest. In Fig. 1.6 reflectance curves of magnesium oxide (commonly accepted as the standard for whiteness) and chemically-bleached cotton are compared; the light reflected by the cotton is deficient in blue and therefore appears yellowish.

When a blue dye is added, a more or less uniform reflectance is obtained throughout the spectrum, but there is a lowering of total reflectance. In effect, blueing decreases the yellowness but makes the cloth darker, rendering the colour of the fabric grey.

FWAs are dyes which absorb ultra-violet radiant energy (always present in daylight) and re-emit the energy as blue light. Since the light emitted by the FWA is added to the reflected light from the cloth, there is an overall increase in reflectance. Thus, the FWA eliminates the yellow cast of the original cloth, but it also increases the amount of light emitted (Fig. 1.6).

Fluorescence

Every "white light" source emits a proportion of rays of wavelength shorter than those of visible light. Some 40 per

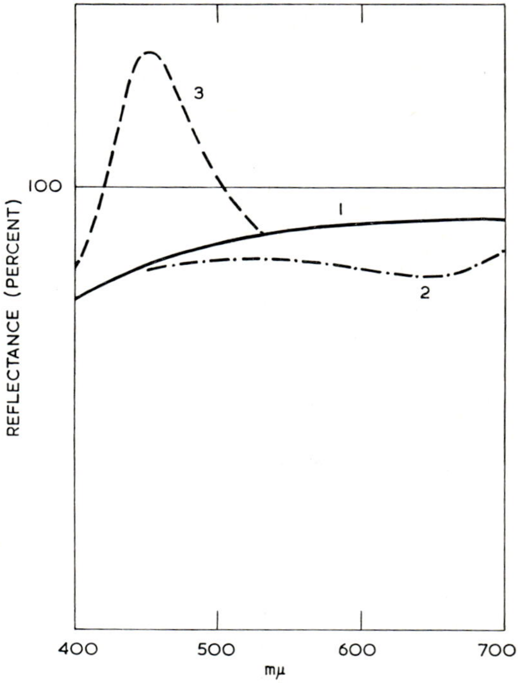

Fig. 1.6. Reflectance of light falling on a substrate. (1) Bleached cotton; (2) bleached cotton treated with "laundry blue"; (3) bleached cotton treated with FWA

cent of the sun's radiation reaches the earth as visible light, 5 per cent as invisible UV and the rest as long-wave infra-red radiation. The rays of shorter wavelength, which are rich in energy, are absorbed partly or wholly by substances on which they fall. The energy levels of the molecules of these substances are increased, the molecules acquiring an "excited state". When the molecules return to their original "ground state", the extra energy is emitted as heat or light.

In the case of non-fluorescent substances, all the energy is emitted as heat within a period of about 10^{-12} to 10^{-13} sec. In the case of fluorescent substances, however, the molecules become electronically and vibrationally excited by absorbing light energy, and are then shielded in some manner so that they lose only their vibrational energy as heat. The molecules then return to their ground state with the emission of the energy represented by electronic excitation as fluorescent light.

The amount of energy emitted as fluorescent light is less than that of the exciting light, the difference normally being equal to the amount of vibrational energy which has been emitted as heat. The quantum of light emitted by an electron when returning to its ground level is generally a smaller energy package than the quantum that is absorbed; the emitted light has a lower frequency and longer wavelength according to Planck's relationship, $E = hf$ (Stokes' Law, proposed in 1852).

Fluorescent materials emit their light over a very short period of time, 10^{-8} sec, the emission following an exponential pattern similar to that of phosphorescent materials. Unlike phosphorescent materials, fluorescent materials do not display an afterglow; they emit light only as long as they are excited.

The use of FWAs depends on the wavelength-shift noted by Stokes; it is concerned mainly with those substances which are essentially colourless or give a colourless solution and which absorb the near-UV radiations of daylight, converting them into blue to blue-violet light.

Fluorescence and Chemical Structure

The fluorescence exhibited by a molecule depends entirely upon its structure and upon the environment in which the spectrum is measured. A non-fluorescent molecule may be converted into a fluorescent one by appropriate chemical treatment. Very few aliphatic or saturated cyclic organic compounds fluoresce. In these compounds the electrons are either very tightly bound or are involved in σ bonding. Absorption of UV energy by a saturated molecule results mostly in bond dissociation because (*a*) high energies are required for excitation of the molecule, and (*b*) electrons are strongly bound.

Energy dissipated in the scission of chemical bonds cannot reappear as fluorescence. Useful fluorescence is restricted to compounds possessing large conjugated systems, containing π electrons. These are less-strongly bound than σ electrons; they can be excited by absorption of radiations of fairly low energy, without extensive disruption of bonding. In aromatic compounds and compounds with conjugated double bonds, therefore, the molecules lose their energy less readily. In liquid solution, most aromatic hydrocarbons exhibit intense fluorescence. Benzene and naphthalene fluoresce in the UV region; anthrancene and naphthacene exhibit violet and green fluorescence respectively (see Table 1.1).

TABLE 1.1

Fluorescence of Aromatic Hydrocarbons (Clar)[11]

Substance	*Colour*	*Fluorescence*
Benzene	Colourless	UV
Naphthalene	Colourless	UV
Anthracene	Colourless	Violet
Phenanthrene	Colourless	UV
Naphthacene	Orange	Green
Chrysene	Colourless	Violet
Pyrene	Colourless	Blue
Triphenylene	Colourless	Blue
Pentacene	Violet	Red

Ring closure often favours fluorescence in aromatic compounds. The most intensely fluorescent aromatic molecules are characterized by rigid planar structure; increasing molecular rigidity decreases vibrational amplitudes. Aromatic molecules with rigid but non-planar structures often display unusually intense emission[12]. The introduction of double bonds conjugated with aromatic nuclei has a profound effect on the fluorescence of molecules. This is illustrated by

TABLE 1.2

Fluorescence of Diphenylpolyenes in Benzene

(Hausser and R. and E. Kuhn)[13]

$H_5C_6—(CH{=}CH)n—C_6H_5$

Number of aliphatic double bonds	*Colour*	*Fluorescence*
1	Colourless	Violet
2	Colourless	Bluish violet
3	Colourless	Blue
4	Yellow	Yellowish green
5	Orange	Yellow
6	Orange	Reddish yellow
7	Orange	Reddish
Tetraphenyl-*p*-xylylene $(H_5C_6)_2.C{=}C_6H_4{=}C(C_6H_5)_2$	Orange	Yellow[14]
p-Distyryl benzene $H_5C_6—CH{=}CH—C_6H_4—CH{=}CH—C_6H_5$		Bluish violet[15]

Table 1.2 which shows the fluorescence of diphenylpolyenes, of which stilbene is a notable example.

Many compounds of the type shown in Table 1.2 exhibit *cis-trans* isomerism. Often, *cis*-isomers are non-fluorescent, whereas the corresponding *trans*-isomers exhibit intense fluorescence, as in the case of stilbene[16]. Occasionally, both the *cis*- and *trans*-isomers fluoresce, but with different energy distributions and efficiencies, e.g. 2-vinylanthracene[17].

A number of fairly reliable generalizations can be made regarding the effect of substituents on the fluorescence of aromatic compounds. Groups which increase the electron density appear to favour fluorescence, whereas groups acting in the reverse way inhibit it. The fluorescence of benzene is shifted to longer wavelengths by substitution (see Table 1.3).

Amino, hydroxy, alkoxy, alkyl and cyano derivatives of benzene show enhanced fluorescence, whereas anilinium, nitro-, azo- and iodobenzene do not fluoresce at all. In the

TABLE 1.3

Fluorescence of Substituted Benzene in Ethyl Alchol

(Ley and Engelhardt)[18]

Substance	*Formula*	*Approximate wavelength of the fluorescence in Å*	*Order of fluorescence intensity*
Benzene	C_6H_6	2700–3100	10
Toluene	$C_6H_5—CH_3$	2700–3200	17
Propylbenzene	$C_6H_5—C_3H_7$	2700–3200	17
Styrene	$C_6H_5—CH{=}CH_2$	2900–3650	14
Phenylacetylene	$C_6H_5—C{\equiv}CH$	2800–4100	10
Fluorobenzene	$C_6H_5—F$	2700–3200	10
Chlorobenzene	$C_6H_5—Cl$	2750–3450	7
Bromobenzene	$C_6H_5—Br$	2900–3800	5
Iodobenzene	$C_6H_5—I$	—	0
Phenol	$C_6H_5—OH$	2850–3650	18
Phenolate	$C_6H_5—O^-$	3100–4000	10
Anisole	$C_6H_5—OCH_3$	2800–3450	20
Aniline	$C_6H_5—NH_2$	3100–4050	20
Anilinium	$C_6H_5—NH_3^+$	—	0
Dimethylaniline	$C_6H_5—N(CH_3)_2$	3250–4050	10
Benzoic acid	$C_6H_5—COOH$	3100–3900	3
Benzonitrile	$C_6H_5—C{\equiv}N$	2800–3600	20
Nitrobenzene	$C_6H_5—NO_2$	—	0

case of conjugated molecules, the fluorescing structures are coplanar with respect to the chromophore.

The absorption spectra of molecules commonly have a direct relationship with their emission spectra in the visible region; there is a mirror-image symmetry between them, as illustrated in Fig. 1.7[19]. In the search for new FWAs, absorption and fluorescence spectra will indicate the most profitable lines to follow, but they will seldom establish the suitability or otherwise of a molecule for use as an FWA. Ultimately, application of the sample to a substrate is the final test of its usefulness.

Most FWAs have absorption maxima around 340–380 mμ and emission around 430–460 mμ; they display only insignificant emissions beyond 500 mμ. Absorption and emission spectra and problems related to hue of emission are discussed in greater detail in Chapter 4.

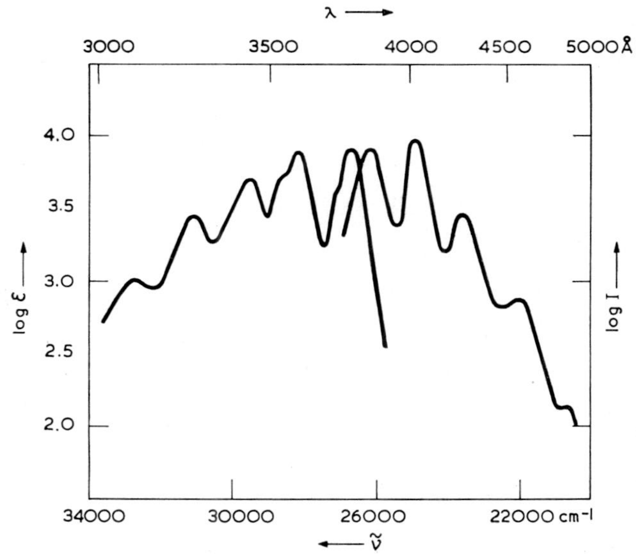

Fig. 1.7. Absorption and fluorescence spectrum of anthracene in dioxane[19]

After Kortnum and Finckh 1942

2

2

CHEMISTRY OF FWAs

The fluorescence and whitening characteristics of many aromatic and heterocyclic systems have been studied in recent years, and the results are to be found in a number of patents and reviews[20]. A few of these compounds have achieved commercial importance, and the chemistry of some of them is considered in this chapter.

1. Derivatives of 4:4′-Diaminostilbene-2:2′-Disulphonic Acid

Most FWAs on the market are derivatives of 4:4′-diaminostilbene-2:2′-disulphonic acid, sometimes referred to as DAS (Fig. 2.1).

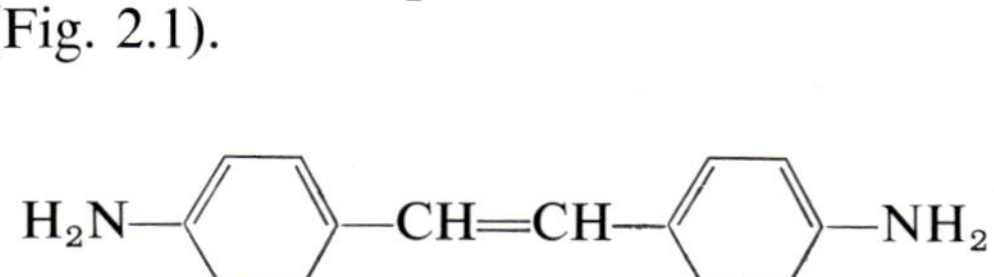

Fig. 2.1. DAS

The simplest derivatives of DAS used as FWAs are *bis*-benzoylamino derivatives[21] (Fig. 2.2), where the benzoyl group may be further substituted by alkoxy[22], hydroxyalkoxy[23], halogenoalkoxy[24] or alkenoxy[25] groups. Alternatively, there may be one alkyl group and an alkoxy or 3:4-methylene-dioxy group[26]; also, the benzoyl group may be replaced by a 3-methoxy-2-naphthoyl group or its tetrahydro derivative[27], or by a phenoxyacetyl group[28]. These compounds, which are no longer in use, are of only

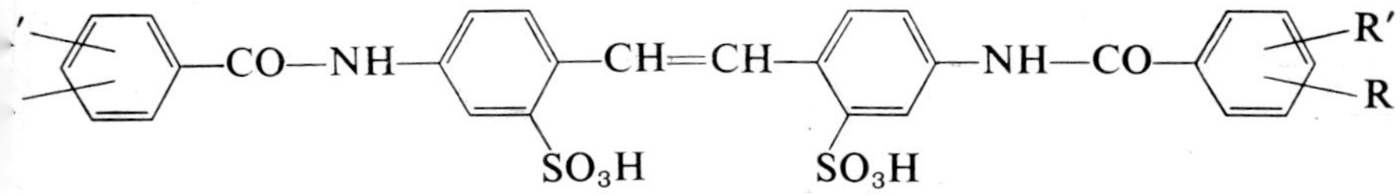

Fig. 2.2

historical importance. They produced a greenish hue on cotton textiles, and even at high concentrations had only a moderate whitening effect. A recent patent describes improved FWAs of this type which are chlorine-resistant; their structure is shown in Fig. 2.3[29].

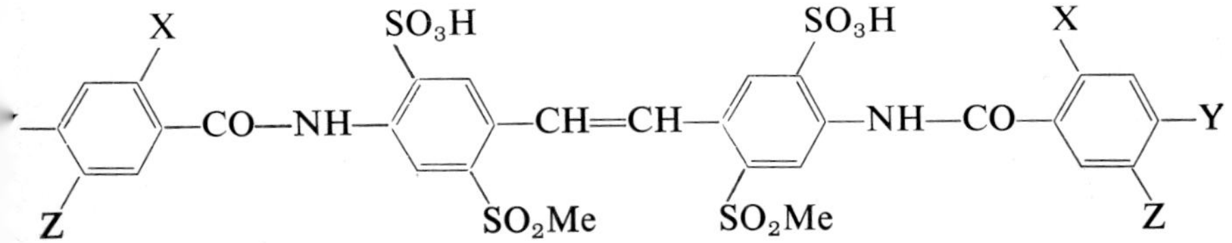

Fig. 2.3

The ureido derivative of DAS[30], (Fig. 2.4), is used in some countries. This compound, first marketed as Blankophor R, is made by the action of phenylisocyanate on DAS. It gives a reddish hue. Blankophor R is unstable to boiling water, which hydrolyses it.

Modifications of the Blankophor R structure have been proposed, such as introduction of halogen, alkoxy, trifluoromethyl groups and other substituents in the phenyl ring[31]. These derivatives do not appear to have gained any more acceptance than the parent compound.

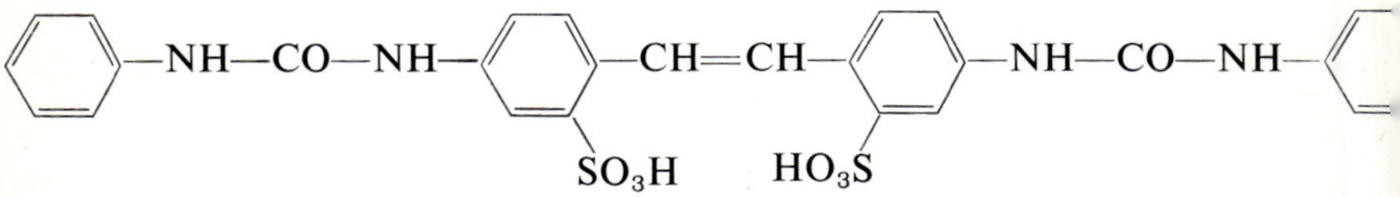

Blankophor R

Fig. 2.4

4:4′-*bis*-Triazinylaminostilbene-2:2′-disulphonic acids—derivatives of DAS—are important FWAs. Blankophor B[30], one of the first of such compounds—no longer in use—was marketed in the early forties; it had the structure shown in Fig. 2.5.

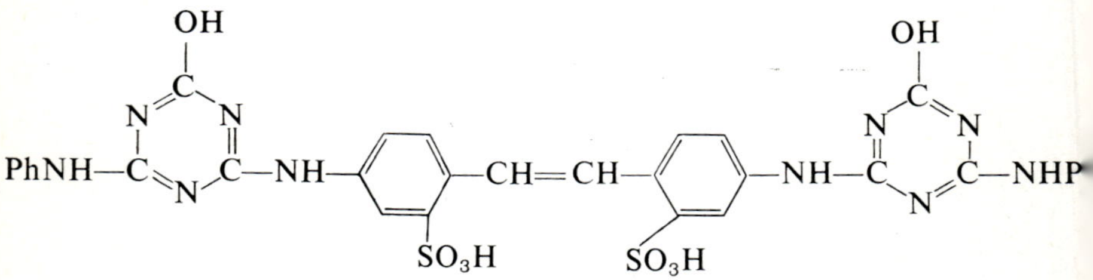

Fig. 2.5

Bis-(triazinylamino)stilbene FWAs are produced as shown in Fig. 2.6.

In Fig. 2.6, X and Y are —NH_2, or the same or different residues of a substituted or unsubstituted primary or secondary aliphatic, alicyclic, aromatic, araliphatic or heterocyclic amine, or OR where R is a substituted or unsubstituted aromatic, alicyclic or aliphatic group.

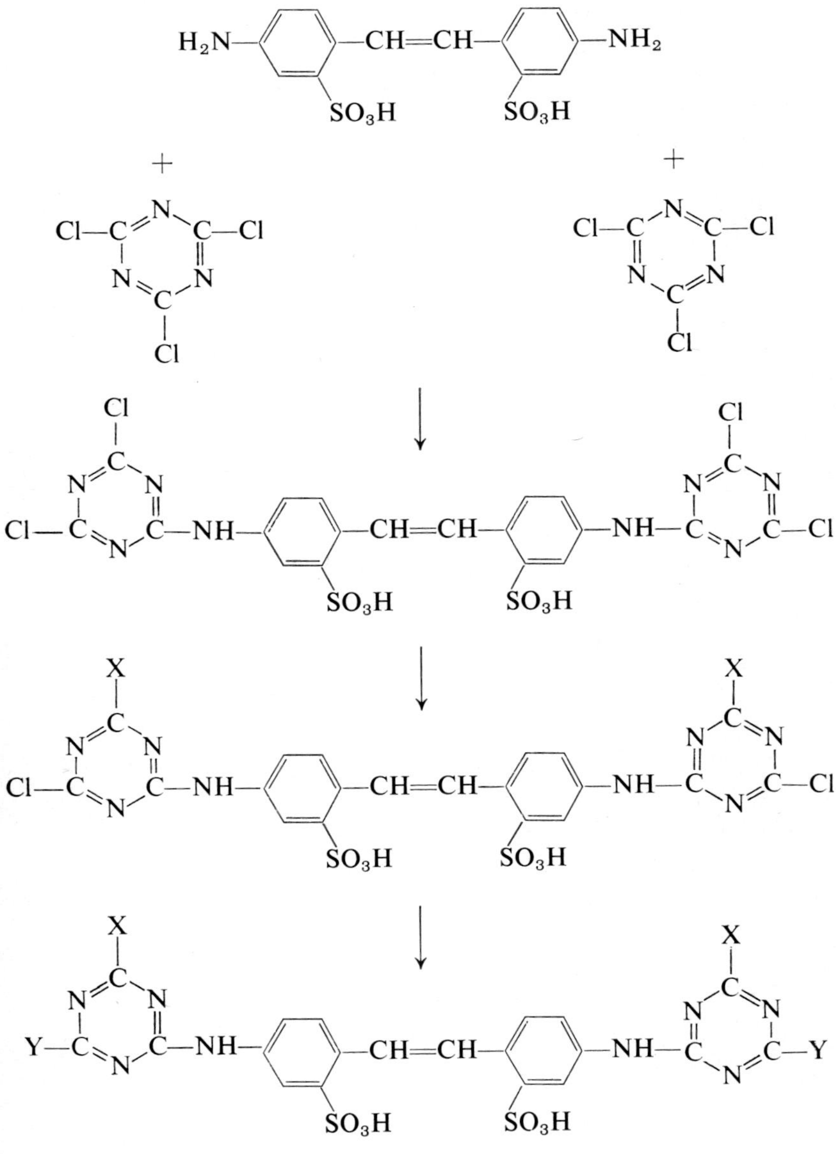

Fig. 2.6. Production of Bis-(triazinylamino)stilbenes

The method of making these compounds depends on the fact that the three chlorine atoms of cyanuric chloride can be replaced in a step-wise manner by the residue of an amine, alcohol or phenol. In general, the first atom in the ring is replaced at temperatures of 0–10°C, the second at 20–50°C and the third at 60–100°C. The reaction temperatures depend upon the nature of the reacting amine[32]. The tetrachloro derivative is not used as an FWA, although claims have been made that it can be applied to cellulosic fibres as a reactive dye[33]. Some dichloro compounds, however, are used for whitening polyamides.

The largest group of compounds in use is that in which both X and Y are residues of an amine. An early patent claims a compound where both X and Y are $—NH_2$ (residue of ammonia), the FWA being obtained by treatment with formaldehyde[34]. This compound is still used.

There are hundreds of patents on *bis*-(triazinylamino) stilbenes covering almost every conceivable combination, but in practice few of these compounds have found extensive use. Because of the relatively low cost of *bis*-(triazinylamino) stilbene disulphonic acid, the price of amines used as X and Y is important. Extra cost must be justified by a significant improvement in the properties of the final compound. The nature of X and Y influences substantivity, solubility, compatibility with different auxiliaries and other properties. Some commercial products are mixtures of two or more compounds; they may be either physical admixtures of individual compounds, or mixtures produced by using a mixture of amines during the synthesis[35].

In recent years, the physical state of some well-established FWAs[36], has been modified, e.g. alteration of crystalline form[37], liquid formulations, etc.[36].

Table 2.1 summarizes some of the important types of *bis*(triazinylamino)stilbene disulphonic acid FWA. No attempt has been made to provide a comprehensive catalogue of patents in this field.

TABLE 2.1

FWAs of the Bis(triazinylamino)stilbene Type

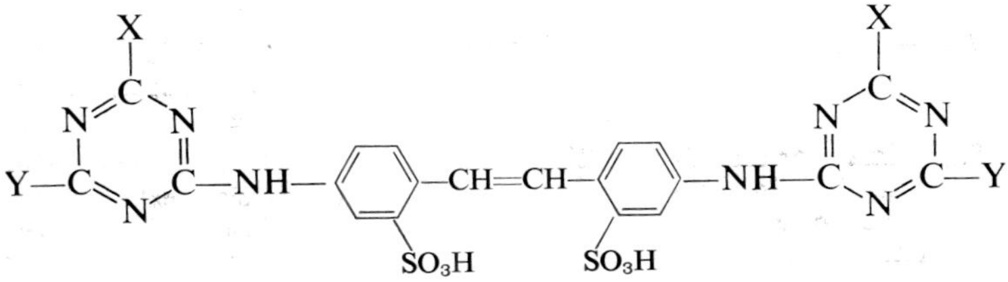

	X	Y	*Application*
1.	Cl	Morpholine[40] or aliphatic amines, e.g. alkanolamines[39], alkylamines[40], substituted or unsubstituted benzylamines[41], diglycolamine[42], alkoxyalkylamines[43], etc.	Particularly for polyamides, also to some extent for cellulosic textiles and paper. (Low affinity)
2.	Cl	Alkoxy[44]	For polyamides, cellulosic textiles and paper. (Low affinity)
3.	Alkoxy (substituted or unsubstituted)	Alkoxy (substituted or unsubstituted)[45]	Paper (for sizing applications)
4.	Alkoxy (substituted or unsubstituted alkyl)	Alkylamine[46]	For polyamides (low affinity) also for cellulosic textiles and paper
5.	Alkoxy	Arylamine[46,47]	High affinity for cellulosic textiles, polyamides and paper
6.	$-NH_2$	Alkylamine or alkanolamine[48]	Particularly suitable for paper, also for textiles. (Low affinity)

(continued)

TABLE 2.1 *(continued)*

	X	Y	*Application*
7.	$—NH_2$	Arylamine[49]	Cellulosic textiles and paper, detergents
8.	Primary or sec. alkylamines incl. alkanolamine, morpholine alkoxyalkylamine, etc.	As X[50]	Low affinity cellulosic textile FWA Also used for paper.
9.	Primary or sec. alkylamines incl. alkanolamine, morpholine alkoxyalkylamine, etc.	Substituted or unsubstituted arylamine[51] (mainly halo, alkyl or alkoxy)	High affinity cellulosic textile FWA—used in detergents, also for paper and polyamides
10.	Primary or sec. alkylamines incl. alkanolamine, morpholine alkoxyalkylamine, etc.	Sulphonatedarylamines[52] (e.g. metanilic or sulphanilic acid)	Cellulosic textiles and paper (not high affinity)
11.	Arylamine	Arylamine[53]	Used in detergent—cellulosic FWA has some affinity for polyamides
12.	Sulphonated[54] alkylamines	Arylamine, alkoxy, alkylamine, etc. or same as X	Paper, cellulosic textiles

Other symmetrical triazinylamino stilbenes[55,56] are shown in Fig. 2.7; they do not appear to have achieved commercial acceptance.

The derivatives of chloroaminostilbene sulphonic acid are claimed to have superior chlorine resistance.

A number of asymmetric triazinylamino stilbenes have featured in the patent literature; they are prepared from 4-amino-4′-nitro-stilbene-2:2′-disulphonic acid. They are of little commercial significance, because (*a*) they are considerably more expensive to prepare than those obtained from

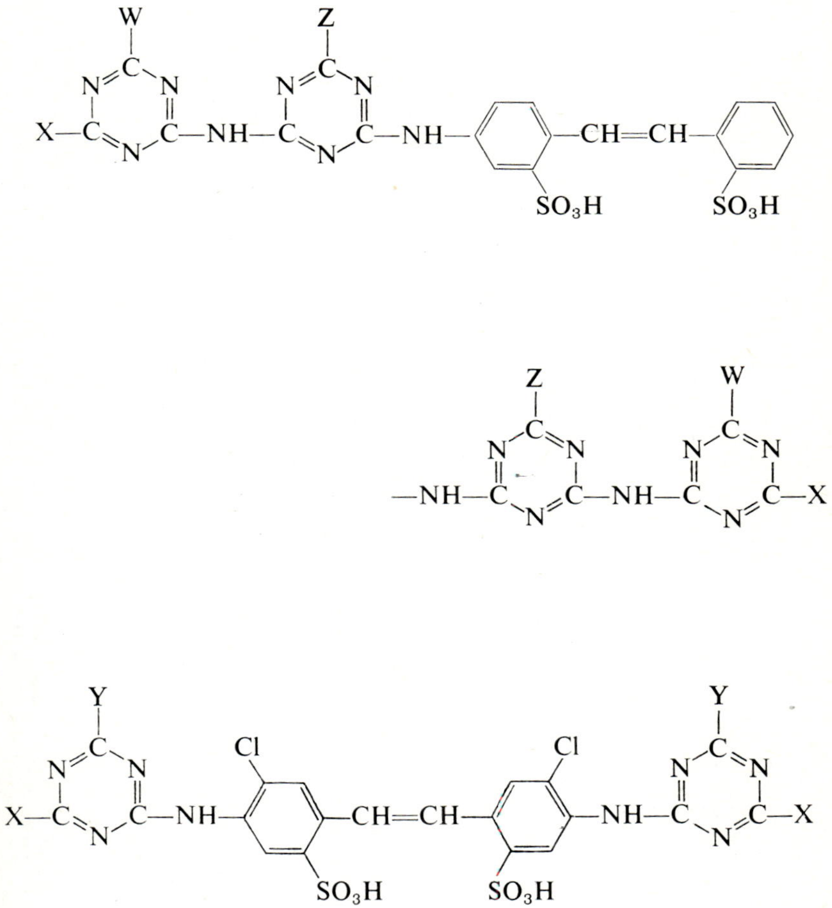

Fig. 2.7

DAS, and (*b*) they show, in general, few advantages over the former. Some of the structures prepared from 4-amino-4′-nitrostilbene-2:2′-disulphonic acid are shown in Fig. 2.8[(57)].

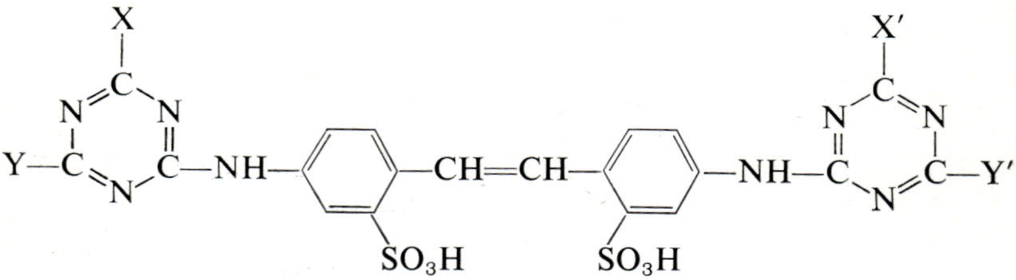

where X, X′, Y, Y′ may all be different

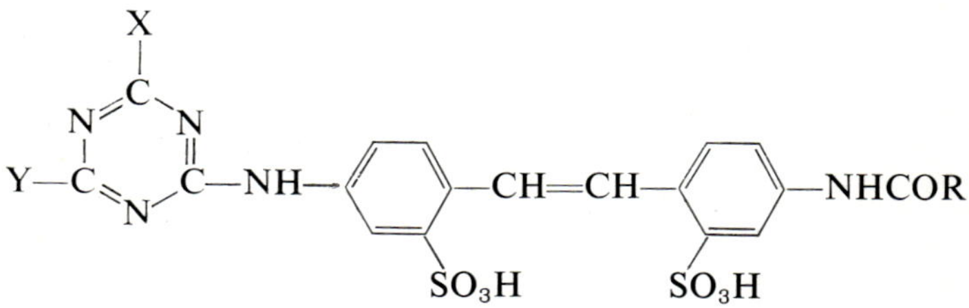

R = substituted or unsubstituted phenyl

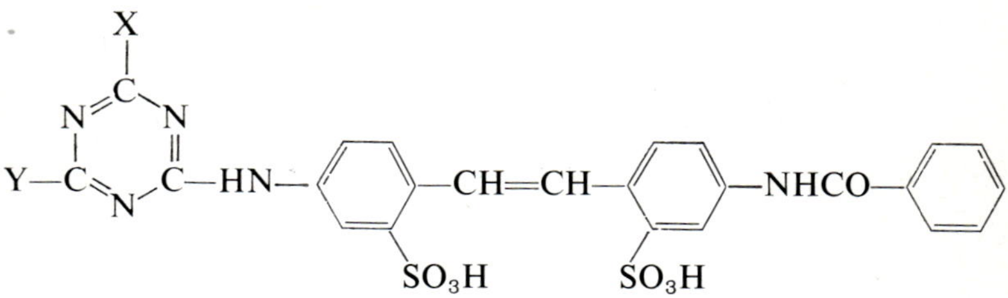

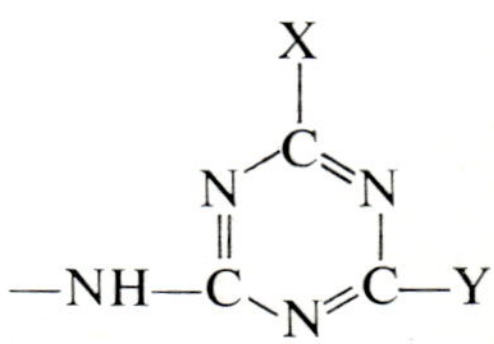

Fig. 2.8 (continued opposite page)

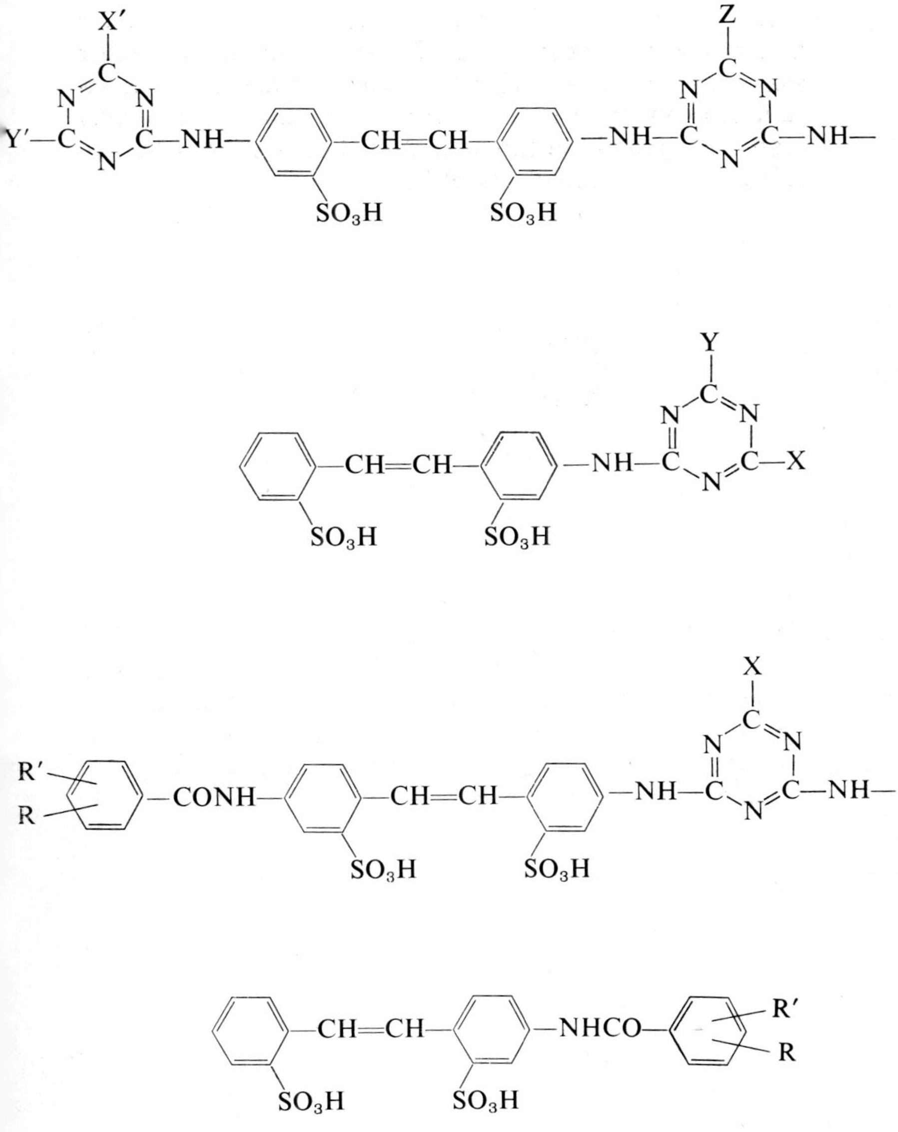

Fig. 2.8 (continuation)

It is claimed that derivatives of stilbene monosulphonic acid of the type shown in Fig. 2.9 give compounds of bluer shade and controlled solubility especially useful for application to polyamide fibres. These compounds are considerably more expensive to prepare than those from 4-amino-4′-nitro-stilbene-2:2′-disulphonic acid(58).

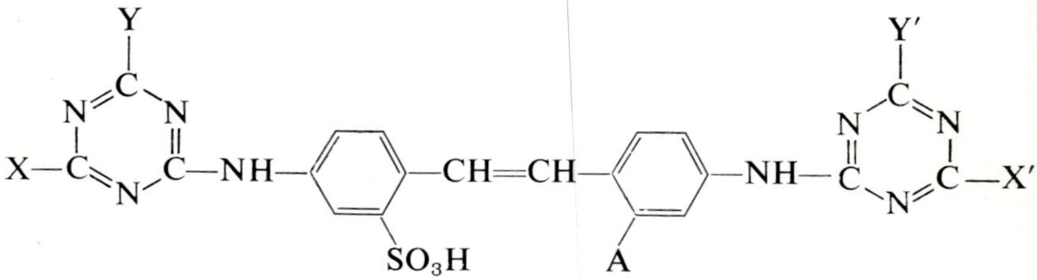

A = H, CH_3, $-OCH_3$, Cl or SO_2NH_2, etc.

Fig. 2.9

2. Styryl Stilbenes and Related Compounds

FWAs with a greenish hue suitable for mixing with FWAs with a violet or red hue are obtained when the conjugation is extended to styryl stilbenes. It is claimed that compounds of the type shown in Fig. 2.10 have good affinity for wool, polyamides and silk(59). Compounds of this type of commercial importance are used for whitening polyester

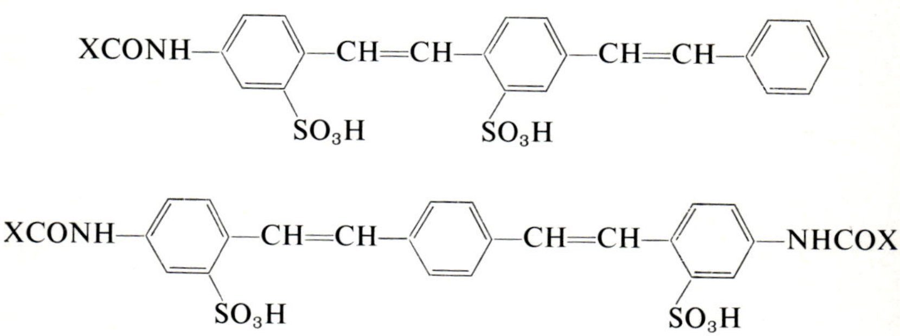

Fig. 2.10

materials[60]; they do not contain sulphonic acid groups or solubilizing groups. They are based on the structure in Fig. 2.11.

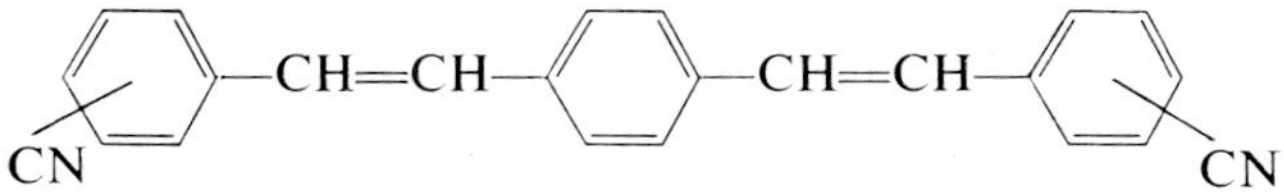

Fig. 2.11

Compounds with terminal rings containing carboxylic acid ester or amide and/or several other groupings which do not impart an appreciable degree of solubility to the final compound provide useful FWAs for synthetic polymeric substances.

Bis-stilbene compounds of the types shown in Fig. 2.12 have attracted attention[61]:

$$R_1—CH{=}CH—X—CH{=}CH—R_2$$

e.g. X = diphenyl residue
R_1 and R_2 = monocyclic benzene diphenyl or naphthalene radical

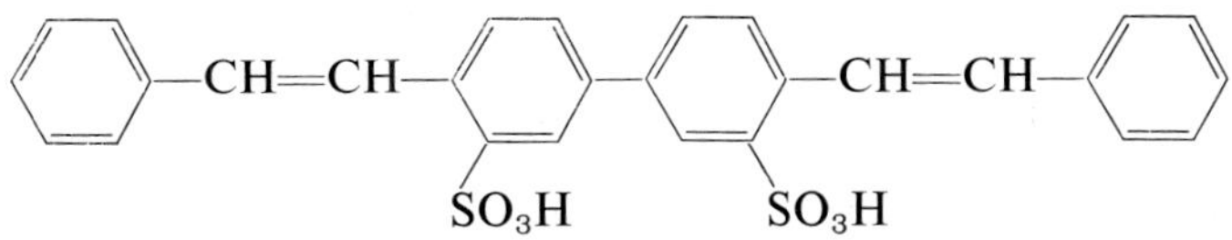

Fig. 2.12

The compound shown in Fig. 2.13 is used for whitening synthetic fibres.

Fig. 2.13

3. **2-Aryl(1:2:3)triazoles**

The *bis*-triazinylamino stilbenes form more than 70 per cent of the total production of FWAs, and under suitable conditions selected members of this class produce a brilliant (violet-blue) white effect with many textile fabrics. These compounds are not, in general, resistant to chlorine-containing bleaching agents like hypochlorite or chlorite. During the 1940s, it was found that *bis*(1:2:3) triazoles obtained by oxidation of *o*-arylamino-azo dyes resulting from the coupling of tetrazotized DAS with β-naphthylamine sulphonic acids were quite stable to chlorine—containing bleaches[62] (Fig. 2.14). Similar compounds from benzidine sulphonic acid and dehydrothio-*p*-toluidine sulphonic acid were also investigated.

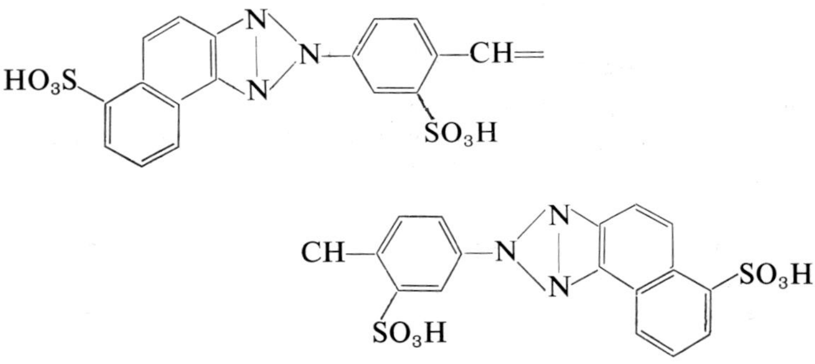

Fig. 2.14

These compounds, which had a pronounced greenish hue, found only limited use, e.g. for neutralizing reddish whitening effects or background contrast effects in certain discharge processes. The compounds possess higher light fastness than the *bis*-triazinylamino stilbenes. Many variants have been investigated in an attempt to improve the hue[63] (Fig. 2.15). None of them has been put to practical use.

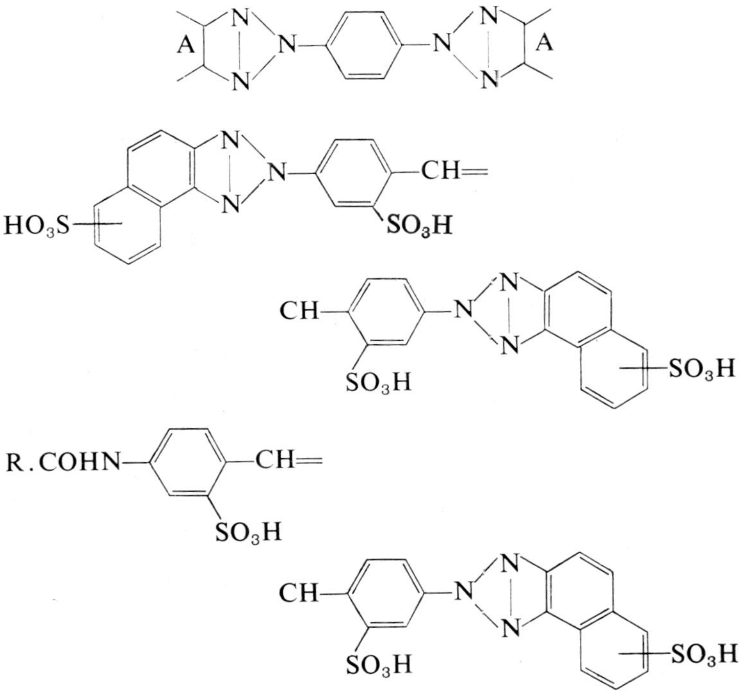

Fig. 2.15

Replacement of the naphthalene residue by an indazole or pyrazole provides FWAs of much bluer or more violet hue(64) (Fig. 2.16).

The type of *bis*-aryltriazole shown in Fig. 2.17 can be used for whitening polyamide and cellulosic materials(65).

Derivatives soluble in organic solvents, used for the whitening of organic polymeric substrates, are obtained by converting the sulphonic acids to esters or amides.

Commercially-important monoaryltriazole FWAs, with neutral to bluish fluorescence, are obtained by substituting 4-amino-stilbene-2-sulphonic acid for DAS(66) (Fig. 2.18).

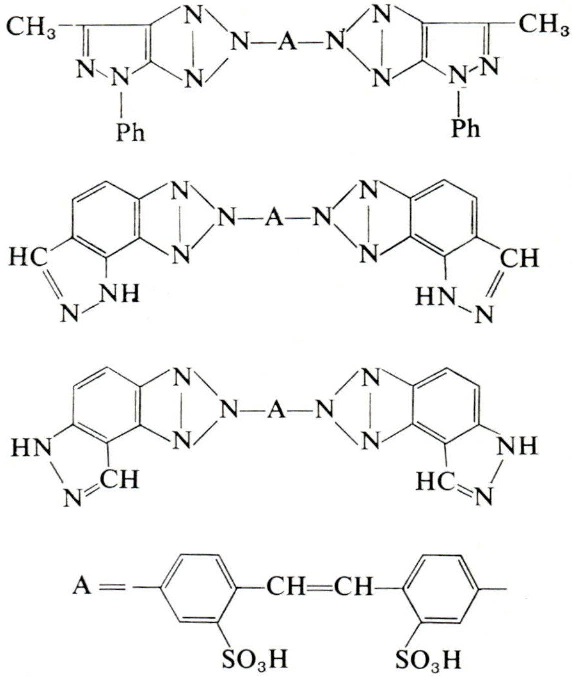

Fig. 2.16

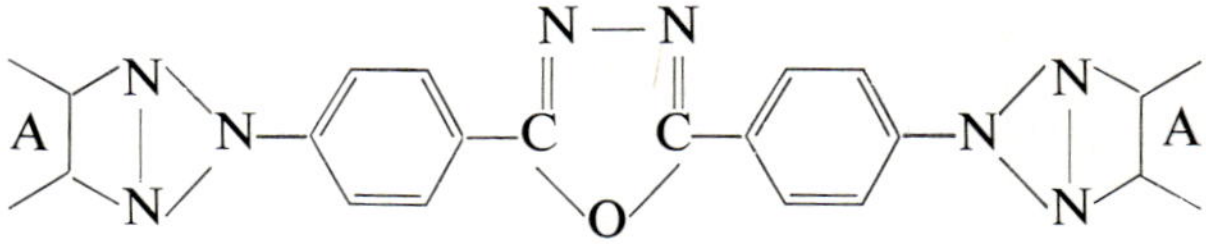

Fig. 2.17

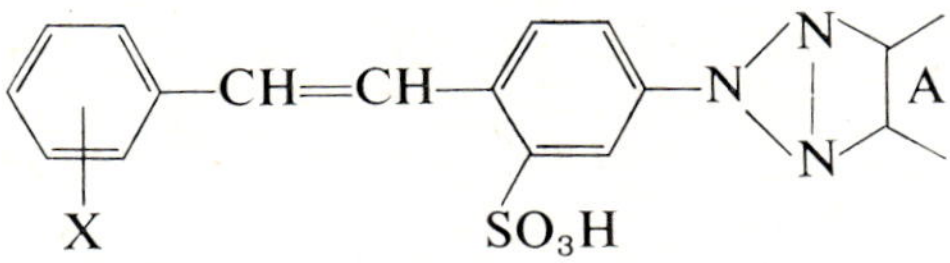

where X = H, Cl, SO_3H etc.
A = benzene, naphthalene, acenaphthene, indazole or pyrazole residue.

Fig. 2.18

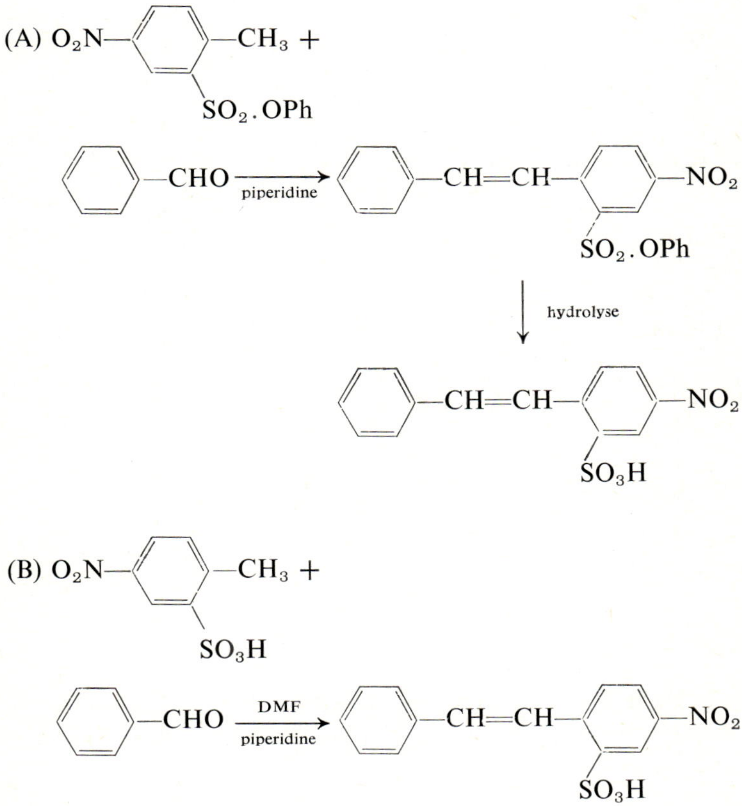

Fig. 2.19

Compounds in which the naphthalene residue is unsubstituted, or is replaced by an acenaphthene residue, are particularly valuable chlorine-resistant FWAs for cellulosic and polyamide fibres, especially when incorporated into heavy-duty detergents. Variants with benzene, indazole, pyrazole residues, etc. have been proposed. None of these compounds, however, is used to any extent for textile finishing (whitening).

Some of the compounds with naphthalene sulphonic acid residue are used in whitening soaps and soap powders. 4-Amino-stilbene-2-sulphonic acid is prepared on a large scale by the reduction of 4-nitro-stilbene-2-sulphonic acid, which in turn is prepared by either of the routes shown in Fig. 2.19[67].

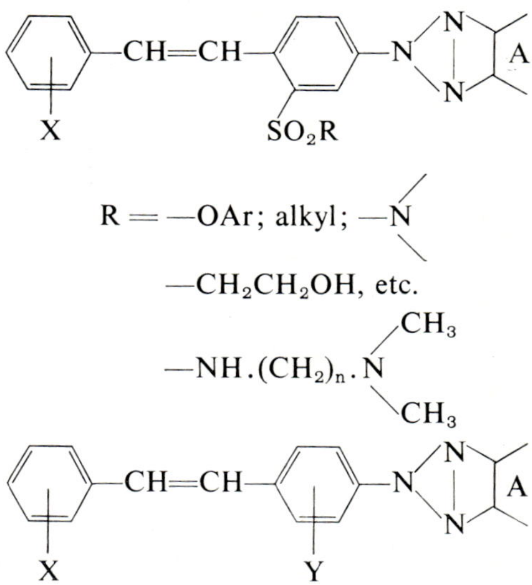

X and Y = cyano, chloro, carboxy ester or amide, etc.

Fig. 2.20

Aryltriazoles obtained from 4-aminostilbene, which do not contain solubilizing groups, have found a variety of uses (Fig. 2.20). Some are used for improving the "colour" of white plastics, oils and other organic substrates. Some are of importance as FWAs for the treatment of synthetic fibrous materials such as polyesters or polyamides[68].

Valuable FWAs derived from the above aryl mono-triazoles include those shown in Fig. 2.21.

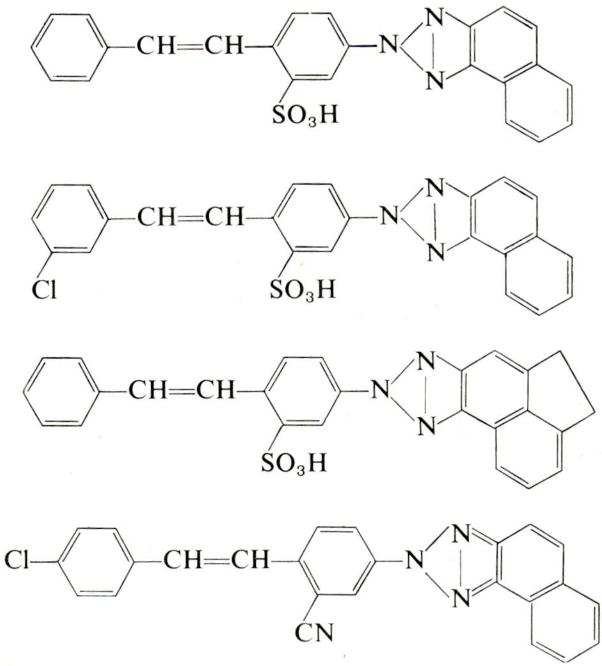

Fig. 2.21

The first three compounds in Fig. 2.21, which contain a sulphonic acid group, are used for whitening polyamide and cellulosic materials from a detergent bath. The last compound is a useful polyester and polyamide brightener.

Early patents[(69)] featured the use of 2(4-aminophenyl)-benzazoles in place of 4-aminostilbene 2-sulphonic acid. In general, these compounds had poor whitening properties compared with the compounds shown in Fig. 2.21. More recently, compounds using heterocyclic analogues of stilbene (Fig. 2.22) have been suggested as useful FWAs for polyamides and cellulose[(70)]; they yield brilliant whites which have too green a hue.

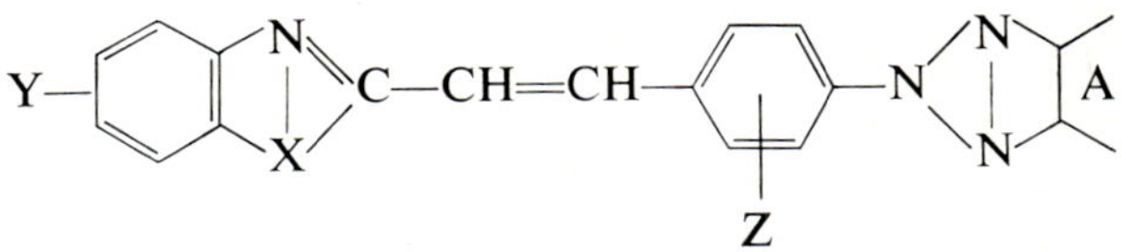

X = O or NH
Y and Z = chloro, alkyl or alkoxy, etc.

Fig. 2.22

4. 1-Aryl-1:2:3-Triazoles and Pyrazoles

The blue to greenish-blue hue of aryltriazoles from stilbenes can be shifted to a more desirable blue to blue-violet fluorescence by the use of

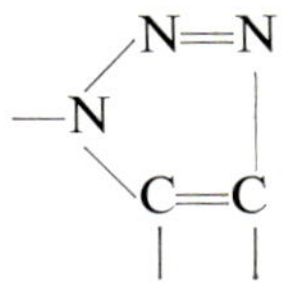

instead of

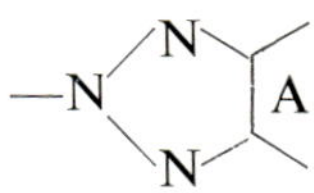

residue (Table 2.2).

TABLE 2.2

(Gold, Petersen and Bayer)[71]

Compound	*Fluorescence*
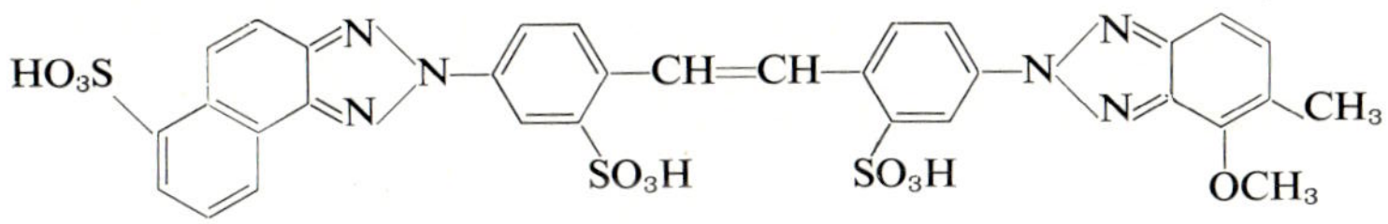	greenish-blue
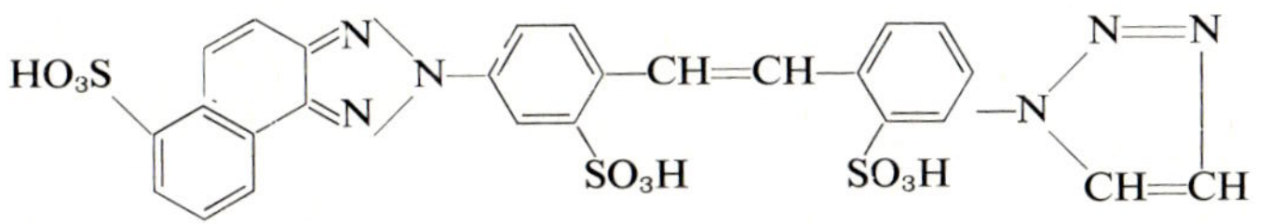 †	blue
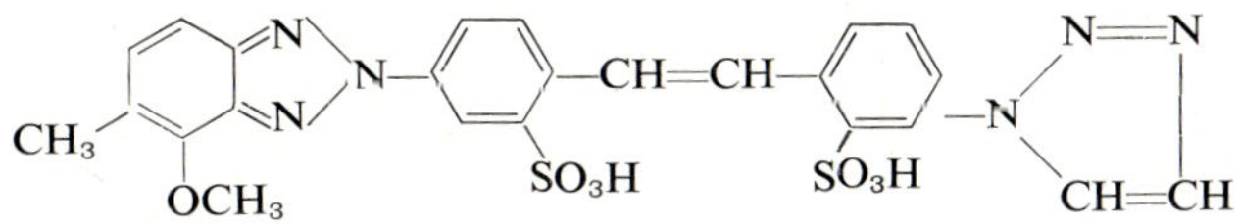	blue
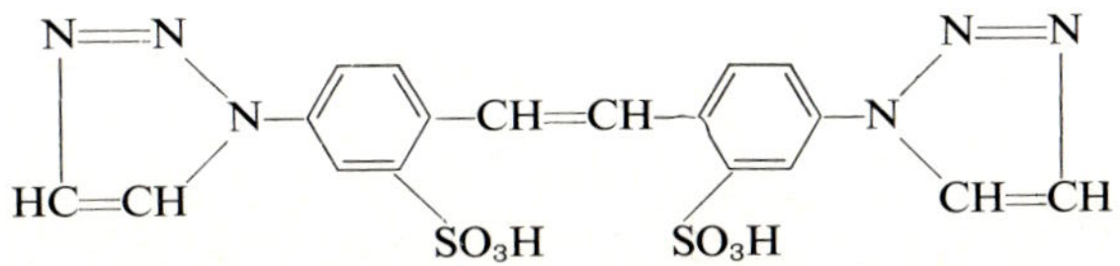	pale violet-blue

† This compound is particularly valuable.

The preparation of 1-aryl 1:2:3-triazoles is carried out as follows:

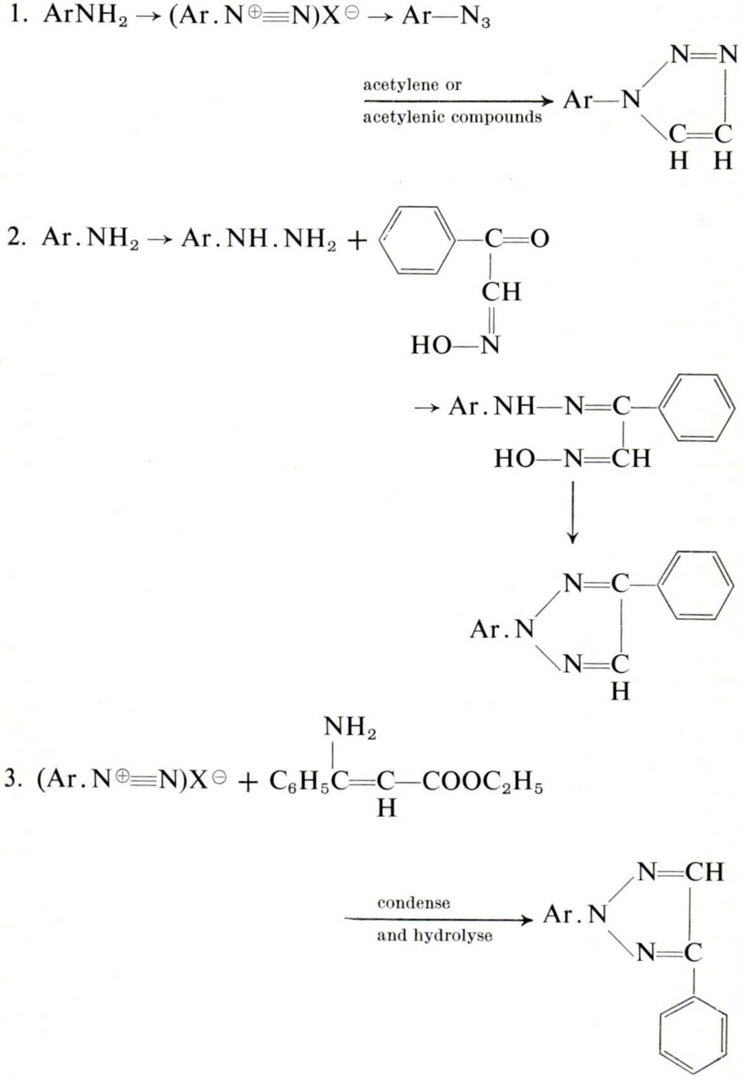

Fig. 2.23

Examples of FWAs of this type are given in Fig. 2.24[72].

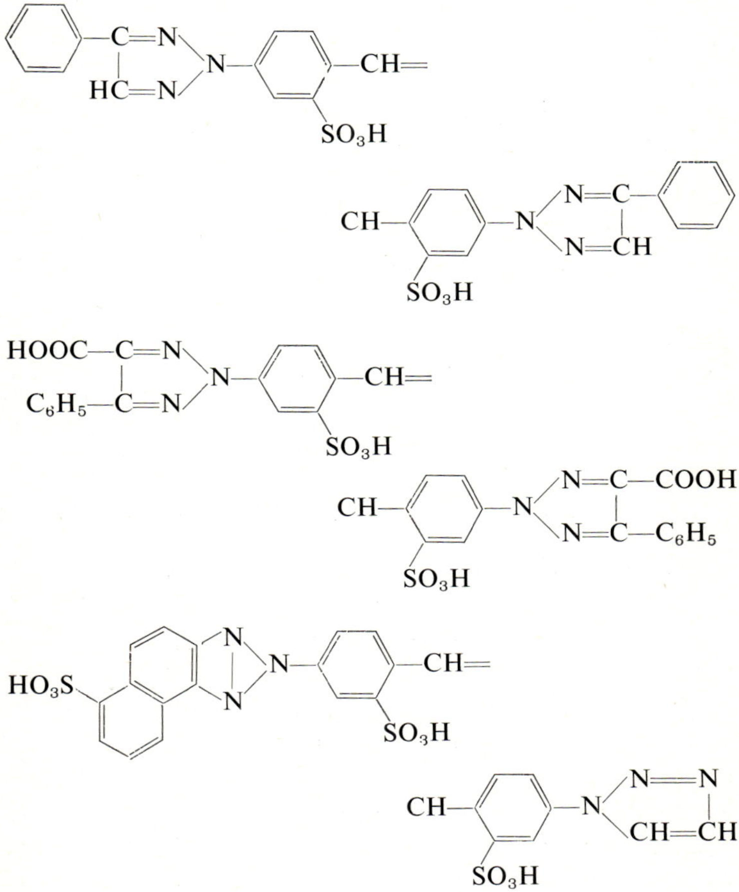

Fig. 2.24

These FWAs can be applied to textiles from a bath containing a chlorine bleach, and in this respect are superior to the *bis*-triazinylamino-stilbene compounds. They do not, however, provide as brilliant a build-up as the latter.

Pyrazole residues 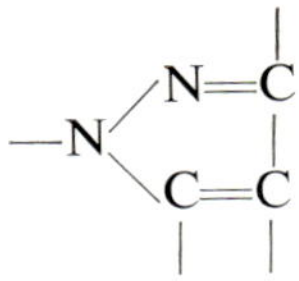 have an effect similar to that of 1:2:3-triazoles, although they are slightly less stable to chlorite than the triazoles(73).

Compounds based on the above principle, using heterocyclic analogues of stilbene, have been investigated(74) (Fig. 2.25). These compounds are non-ionic and have

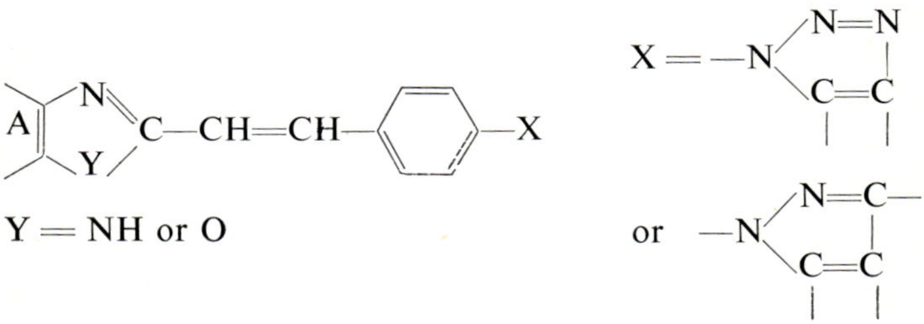

Fig. 2.25

reasonably good affinity for many substrates, including cellulose, polyamides and polyesters. They are resistant to chlorine-containing bleaches and can be applied to cellulose from a bath containing a cationic softening agent.

5. **Benzothiazoles, Benzimidazoles and Benzoxazoles**

The fluorescent properties of dehydrothio-*p*-toluidine, an intermediate for dyestuffs, have been known for a long time. However, the intrinsic yellow colour of the compound makes it useless as a FWA. The alkylated or acylated derivatives of this compound give a useful whitening effect, but the effect is extremely fugitive to light and is of little practical use. Benzimidazoles and benzoxazoles have received considerable attention in the search for FWAs suitable for use with synthetic fibres.

An early patent(75) in this field covered compounds of the general formula:

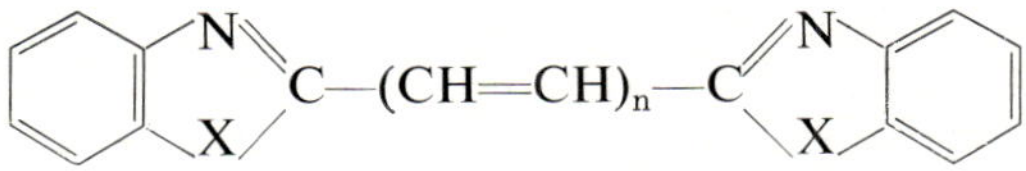

where X = NH, O or S. n = 0 or 1

Fig. 2.26

One of the earliest products marketed had the following structure:

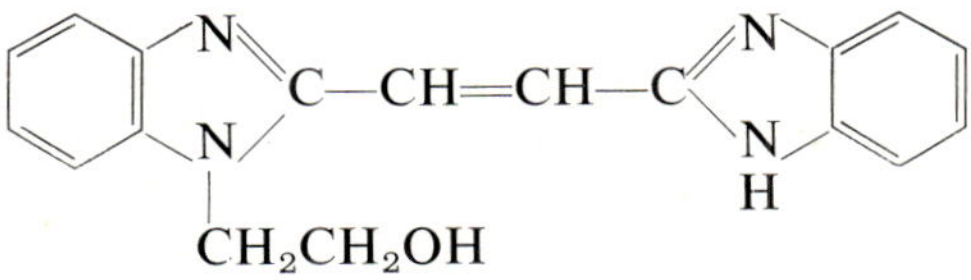

Fig. 2.27

The compound, which yields a greenish-blue hue, has reasonable affinity for both polyamide and cellulosic fibres. It has good resistance to hypochlorite, and is incorporated in household detergent powders.

Many patent applications have been concerned with *bis*- and *mono*-arylimidazoles, some of which have useful properties as FWAs for polyamide and acrylic fibres(76). Patents have also been devoted to improved methods of manufacture of these compounds, examples of which are shown in Fig. 2.28.

Y_1 and Y_2 are hydrogen or substituted or unsubstituted alkyl, alicyclic, aryl, aralkyl, etc.; the benzene rings may also be substituted. Quaternary ammonium compounds derived from some of the structures in Fig. 2.28 are especially useful for whitening acrylic fibres. Some benzimidazoles have affinity for resin-treated cotton.

Some substituted monobenzimidazoles, and quaternary ammonium compounds derived from them, yield valuable FWAs for acrylic fibres(77) (Fig. 2.29).

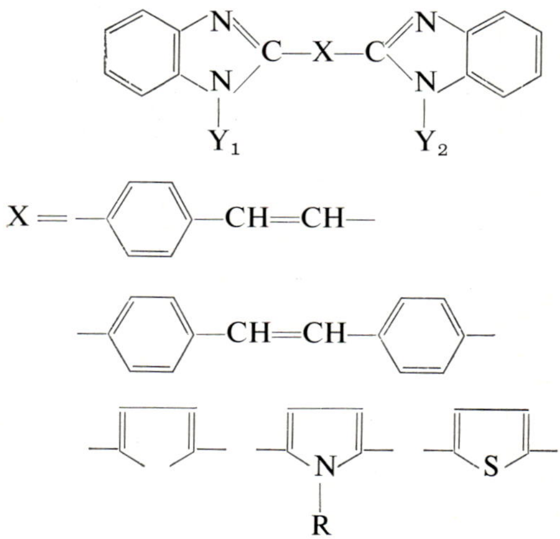

Fig. 2.28

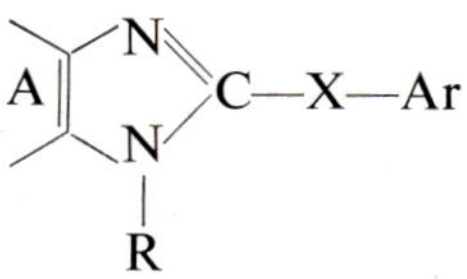

A and Ar = substituted or unsubstituted aromatic radicals
R = aliphatic or aromatic radical

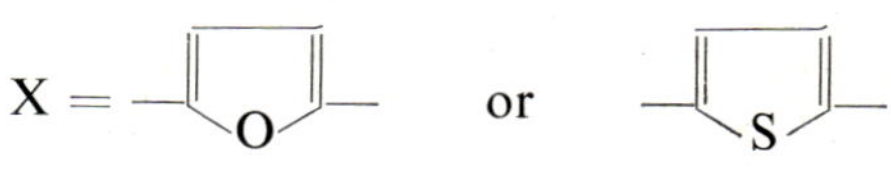

Fig. 2.29

Many important FWAs for polyesters are benzoxazole derivatives. One of the earliest products of this type was:

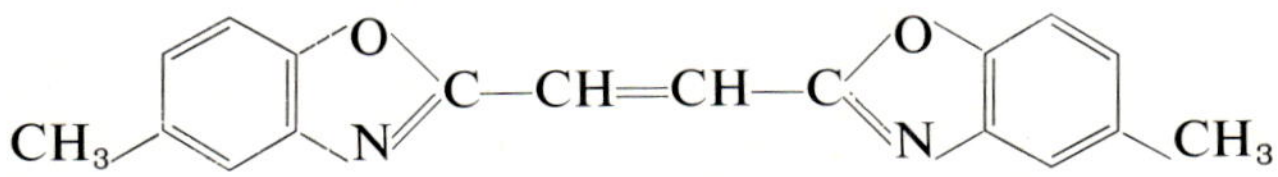

Fig. 2.30

This FWA, first marketed in 1956, is still used extensively for whitening synthetic fibres, particularly polyesters. Many patents have been concerned with the manufacture of this and similar compounds(78), which can be prepared by processes shown in Fig. 2.31.

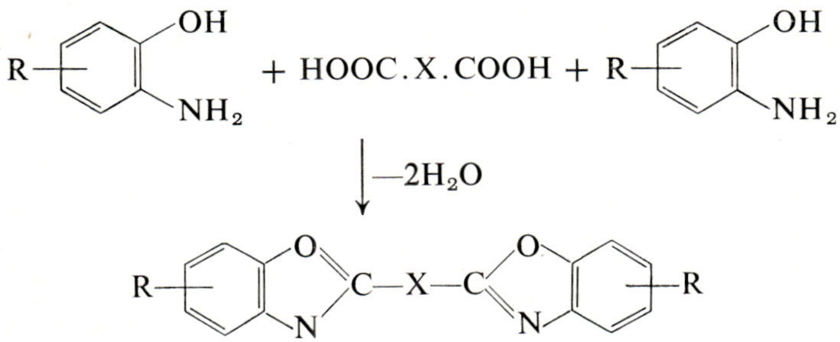

Fig. 2.31

Interesting and effective FWAs are obtained when X (in Fig. 2.31) is

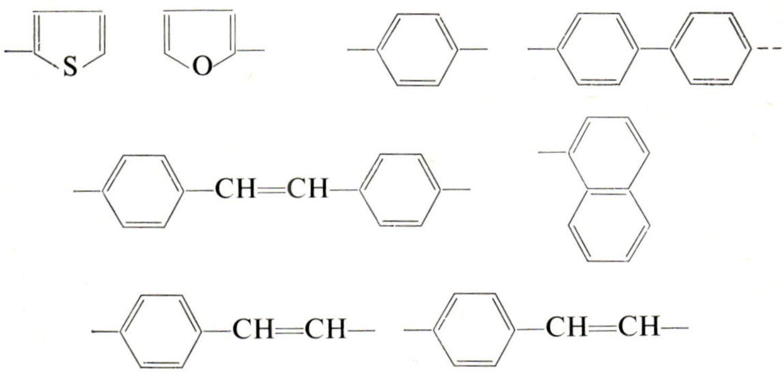

Many of these compounds are particularly suitable for whitening synthetic fibres by incorporation during melt spinning.

Important FWAs for synthetic fibres and polymeric materials are derived from the 2-styryl benz-[80] or naphthoxazoles[81]. Examples are shown in Fig. 2.32.

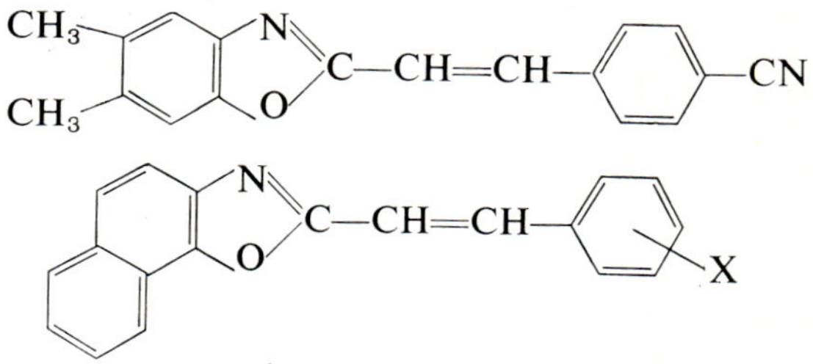

where X = H or Cl

Fig. 2.32

2-Substituted benzoxazoles have been suggested as FWAs to be incorporated in synthetic polymeric materials during melt spinning, including those in Fig. 2.33[82].

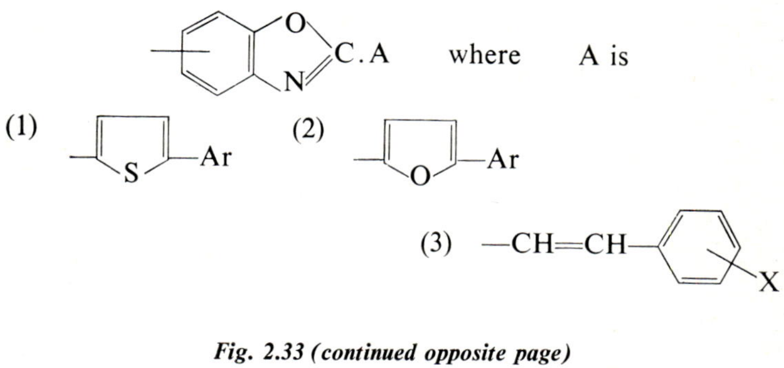

Fig. 2.33 (continued opposite page)

(4) 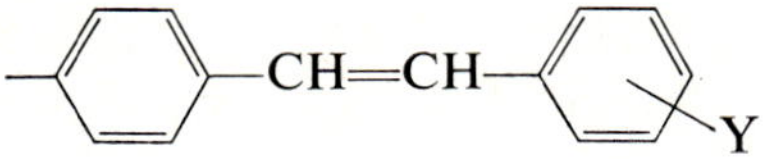

(5) 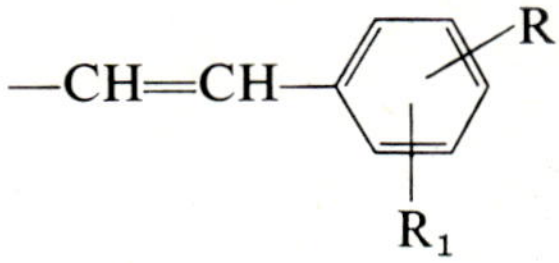

(6) 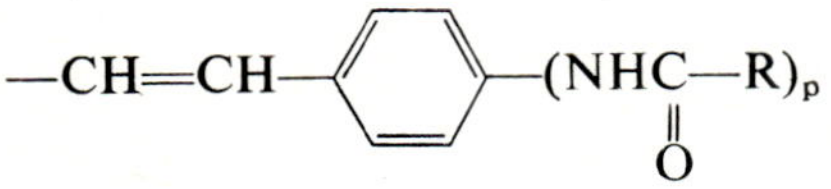

(7) 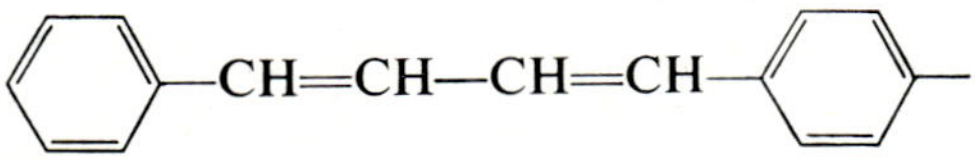

(8) 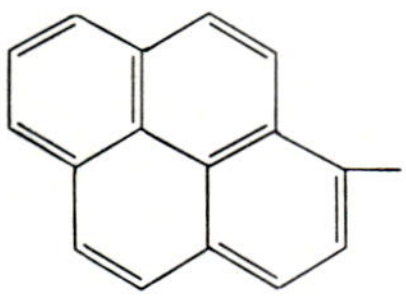

(9) 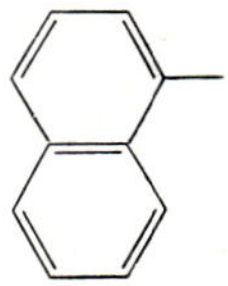

(10) 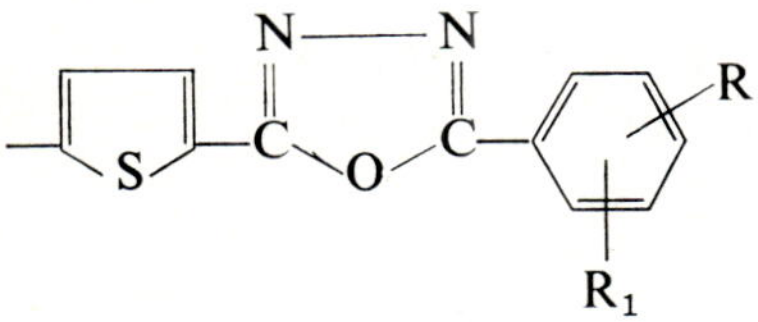

(11) 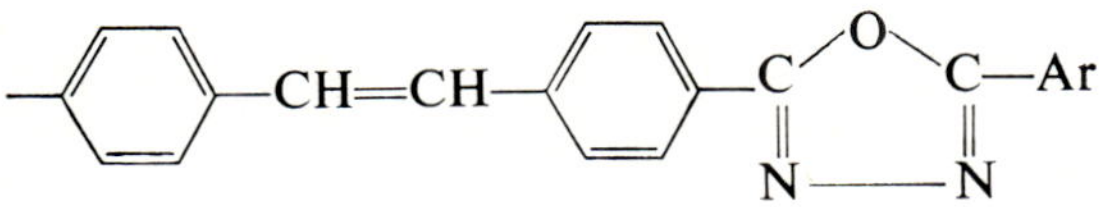

(12) 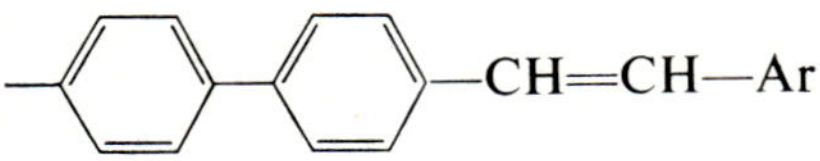

(13) 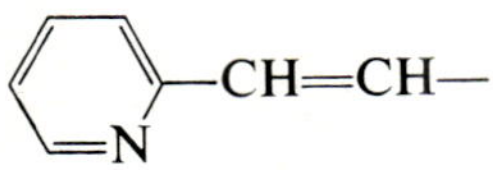

Fig. 2.33 (continuation)

Quaternary ammonium compounds derived from benzoxazoles[83] yield important FWAs for acrylic fibres (Fig. 2.34).

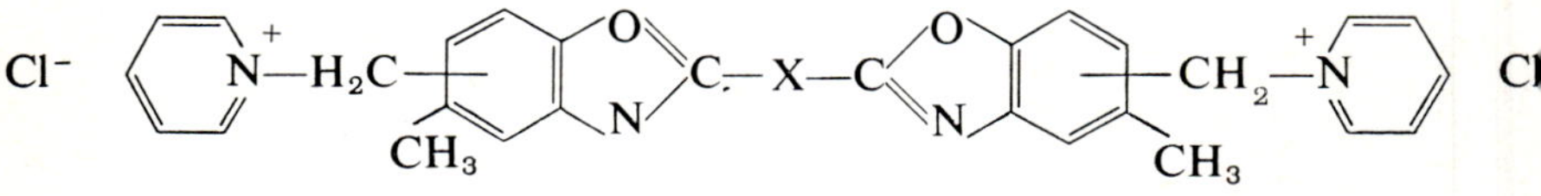

where X = —CH=CH— or 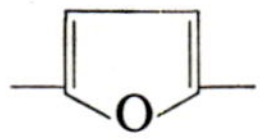

Fig. 2.34

6. **1:3:4-Oxadiazoles**

1:3:4-Oxadiazoles have attracted much attention, many of these compounds being particularly useful for incorporation in polymer melts. Some of the compounds[84] can be represented as in Fig. 2.35

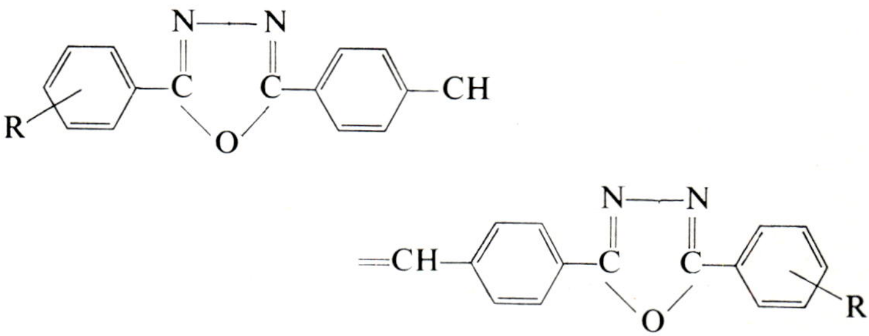

R and R′ can be alkyl, halogen, alkoxy, carboxyamide or ester or cyano

Fig. 2.35 (continued opposite page)

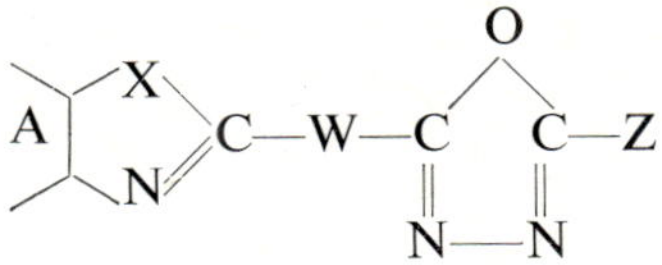

A = benzene, naphthalene or tetrahydronaphthalene
W = stilbene residue
Z = aryl, aralkyl or aryl heterocyclic

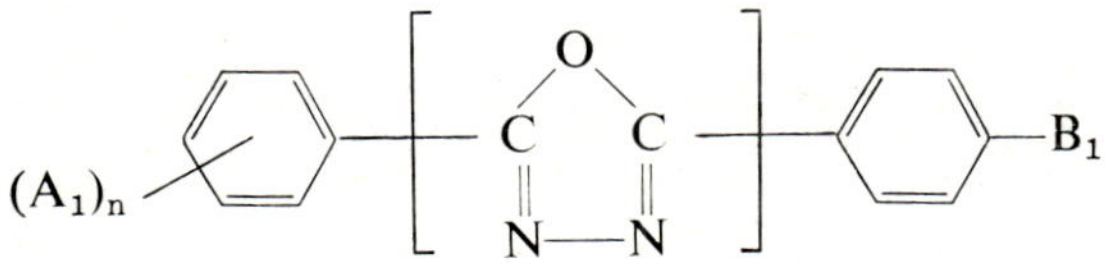

A_1 = C_3—C_7 branched alkyl
m and n = 1 or 2
B_1 = H or phenyl or C_1—C_7 alkyl or C_1—C_4 alkoxy

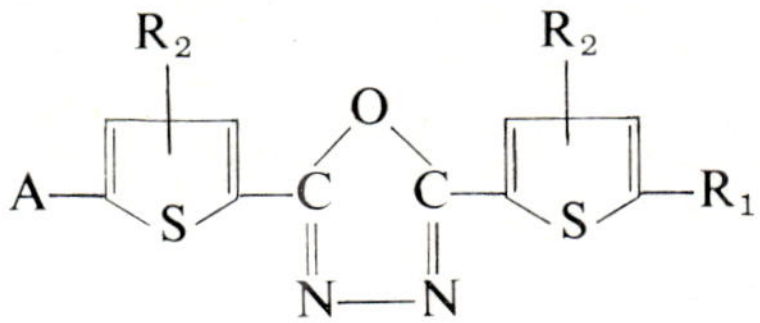

A = phenyl or benzoxazolyl
R_1 = hydrogen or halogen atom, straight or branched chain alkyl, haloalkyl, hydroxyalkyl, cyano alkyl, etc.
R_2 = H, halogen or alkyl.

Fig. 2.35 (continuation)

One of the compounds used in the polyester spinning mass[85] has the following formula:

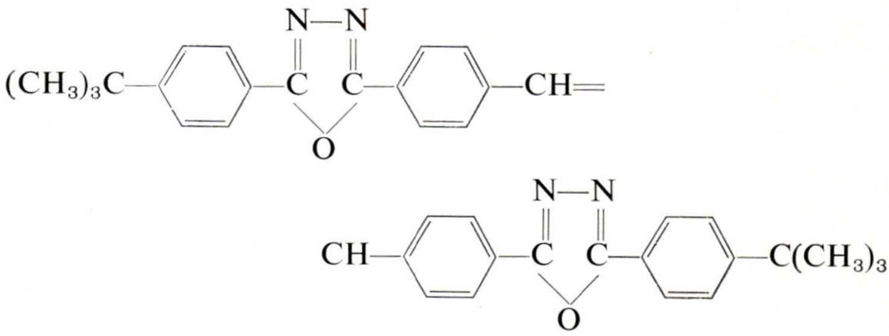

Fig. 2.36

Similar compounds based on thiadiazoles have also been claimed[86].

7. **Coumarins**

β-Methyl umbelliferone, one of the first known whitening agents, is still used to a limited extent for the whitening of soaps. It is no longer used for whitening textiles. The 7-dialkylamino-4-methyl-coumarins, despite their very poor light fastness, are used for whitening wool and some man-made fibres including cellulose acetates and polyamides (Fig. 2.37).

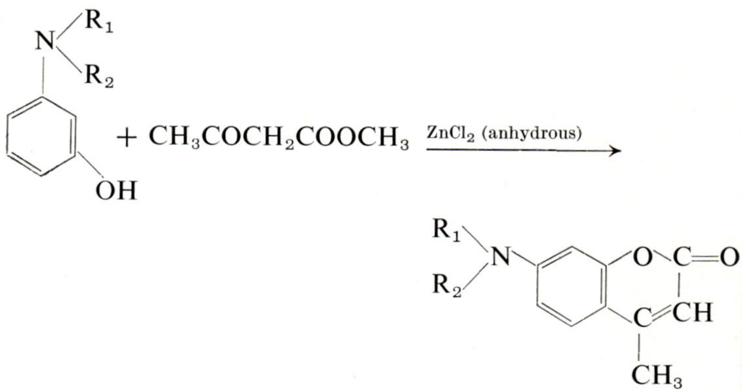

Fig. 2.37

Variants[87] of the above have been investigated, but none has shown any significant advantage over 7-dimethylamino- and 7-diethylamino-4-methyl-coumarin.

During the early fifties it was discovered that derivatives of 7-hydroxy and 7-amino coumarins substituted in the 3-position with a phenyl or a heterocyclic residue give interesting FWAs; some have much higher light fastness than the compounds not so substituted. Simple amino derivatives, such as the acyl or ureido derivatives[88], have not been used to any great extent. Some 7-alkoxy compounds[89] yield useful FWAs for synthetic fibres (Fig. 2.38).

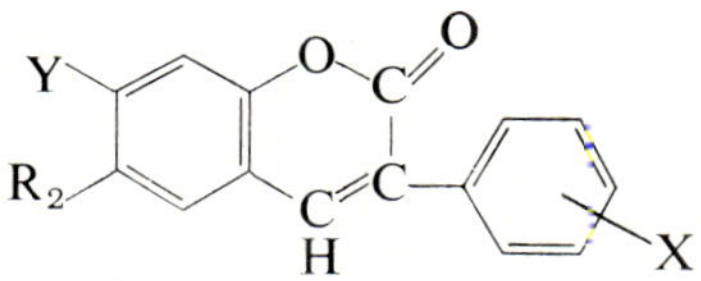

$X = H$ or CN
$Y = —OR_1$ or $—SR_1$
$R_2 = H$ or alkyl

Fig. 2.38

Derivatives of amino coumarins which give useful and interesting FWAs are shown in Fig. 2.39[90].

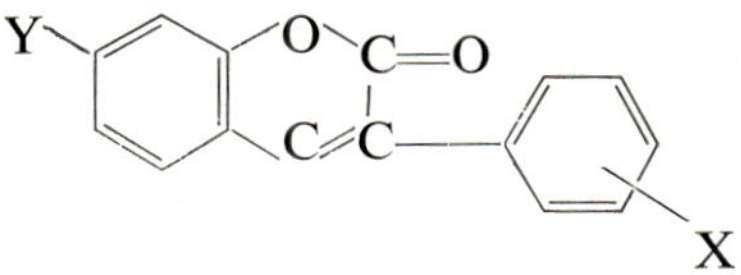

Fig. 23.9 (continued over)

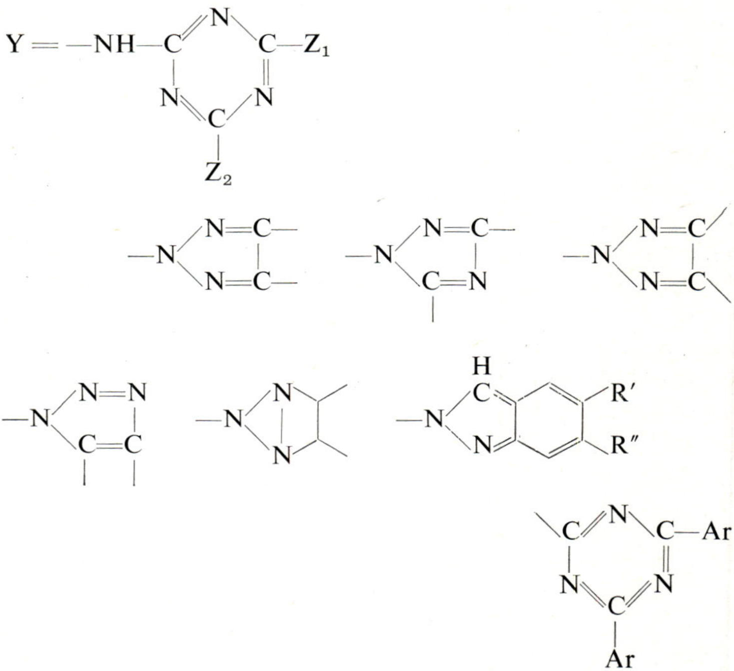

Fig. 2.39 (continuation)

Some of the 3-heterocyclic derivatives covered are as follows:

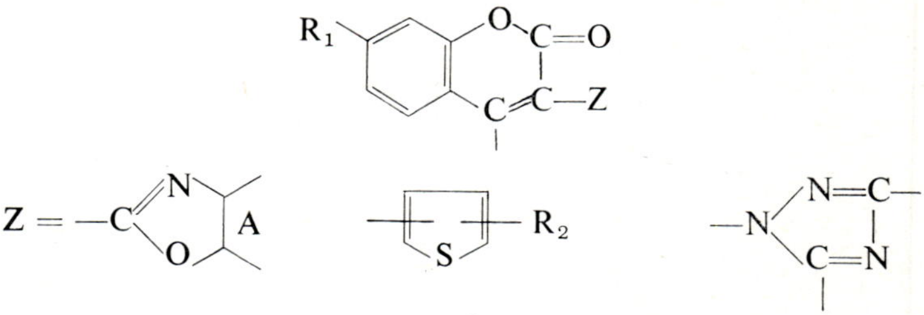

Fig. 2.40

The 7-amino and 7-hydroxy coumarins are prepared by the following sequence of reactions:

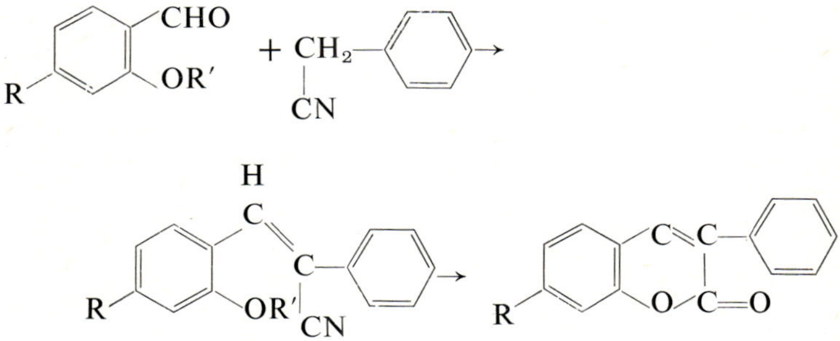

R = —NH—acyl, or —O—acyl

Fig. 2.41

Investigation of carbostyrils[92], which have a structural similarity to coumarins, has yielded some interesting products, e.g.

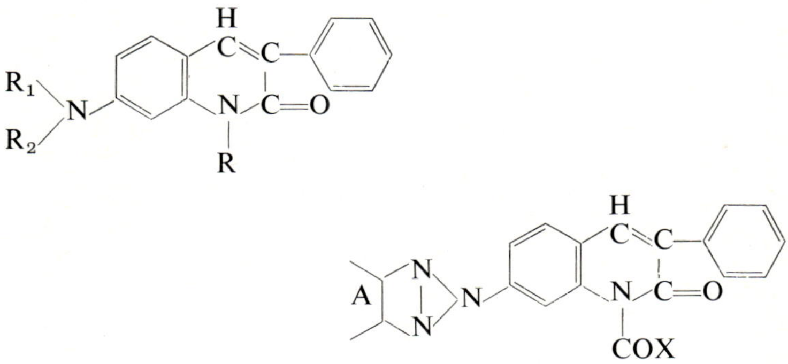

Fig. 2.42

7-Dialkylamino compounds have a particularly good whitening effect on polyamides and polyacrylonitriles.

8. **Pyrazolines**

It has long been known that 1:3-diary land 1:3:5-triaryl pyrazolines are fluorescent. In 1950, Ilford Ltd[93] found that they were useful for whitening textiles. Some water-soluble compounds were used for wool and polyamides, and a number of 1:3-diaryl pyrazolines[94] were subsequently used as FWAs for whitening polyamides and acrylic fibres (Fig. 2.43). They are used extensively for this purpose, especially in Europe.

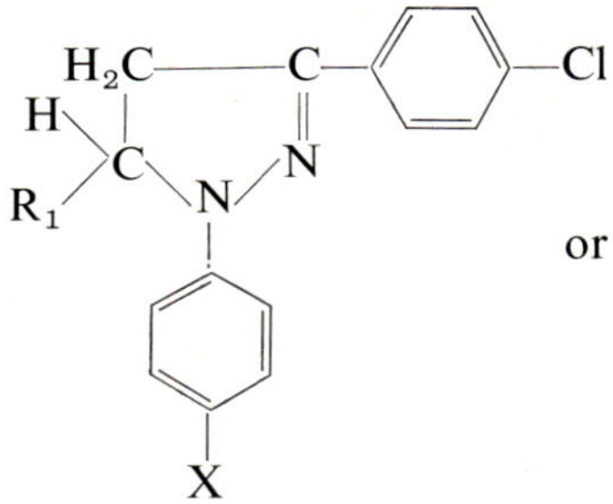

or

$X = -SO_3H,\ -SO_2NH_2,\ -SO_2.CH_2CH_2OH,$
$-SO_2CH_3,\ -SO_2.CH_2CH_2SO_3Na,$
$-SO_2CH_2CH_2NHCOR,$

$$-SO_2.CH_2CH_2.OCH_2CH_2N\begin{matrix}\diagup R_1\\ \diagdown R_2\end{matrix},$$

$$-SO_2.CH_2CH_2N\begin{matrix}\diagup R_1\\ \diagdown R_2\end{matrix}$$

$-COOH$ or $-COOR$

Fig. 2.43 (continued opposite page)

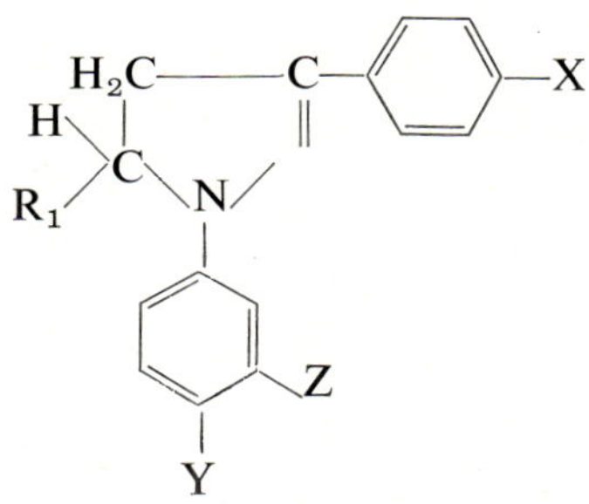

$X = H, Cl$ or $NHCOCH_3$
$Y = H$ or Cl
$Z = Cl$ or CF_3

Fig. 2.43 (continuation)

Preparation and properties of these compounds have been discussed by Petersen *et al.*[95].

9. **Naphthalimides**

Acyl derivatives of 5-aminonaphthalimide (Fig. 2.44) have long been known to possess good whitening properties. Because of their greenish hue, these compounds have not found extensive acceptance.

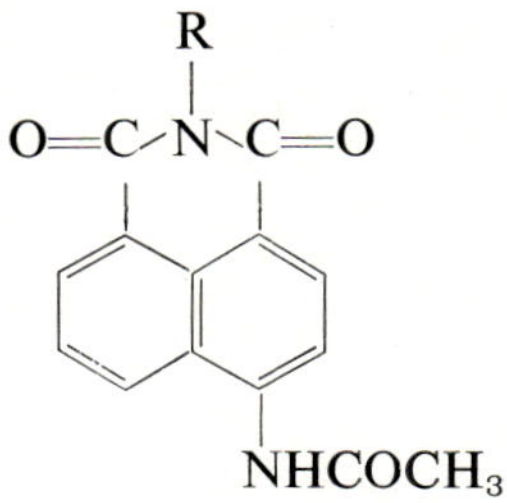

R = alkyl (particularly butyl)

Fig. 2.44

Replacement of the acetamido group by 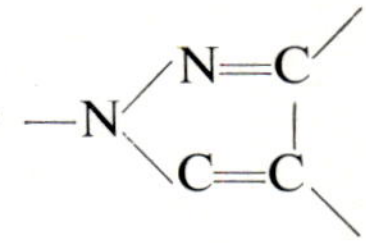 or

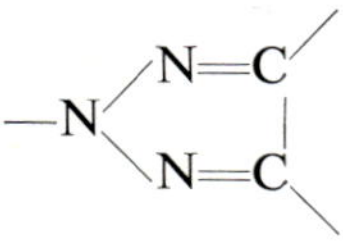 yields FWAs of bluer hue[97].

FWAs of greater importance are obtained when the acetamido group is replaced by a methoxy group[98]. The 5-alkoxy naphthalimides are very versatile FWAs; they can be used for a variety of synthetic fibres, including polyesters, acrylics and polyamides. They are prepared quite readily from 5-nitro-acenaphthene by oxidation and the usual replacement reactions. The corresponding S-alkyl compounds afford FWAs with greenish hue[99].

10. **Derivatives of 2:6-Diamino-dibenzthiophenedioxide-3:7-disulphonic Acid**

Acyl derivatives of 2:6-diaminodibenzthiophenedioxide-3:7-disulphonic acid (Fig. 2.45) possess outstanding stability to chlorine-containing bleaches. They display rather weak fluorescence, however, and tend to build up to a greenish hue on cellulosic fibres. They are useful when resistance to high dosages of chlorine-containing bleaches is required[100].

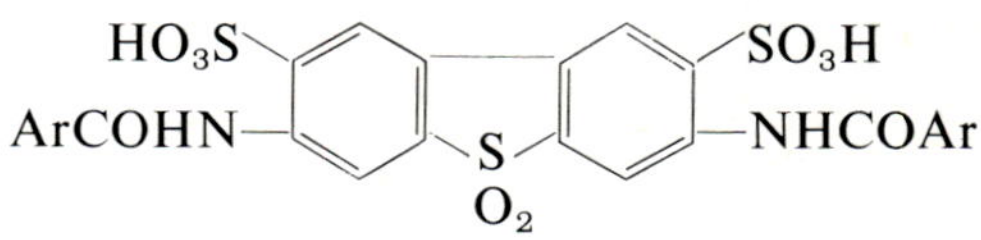

Fig. 2.45

1:2:3-Triazole derivatives of dibenzthiophene dioxide are used for whitening synthetic fibres[101] (Fig. 2.46).

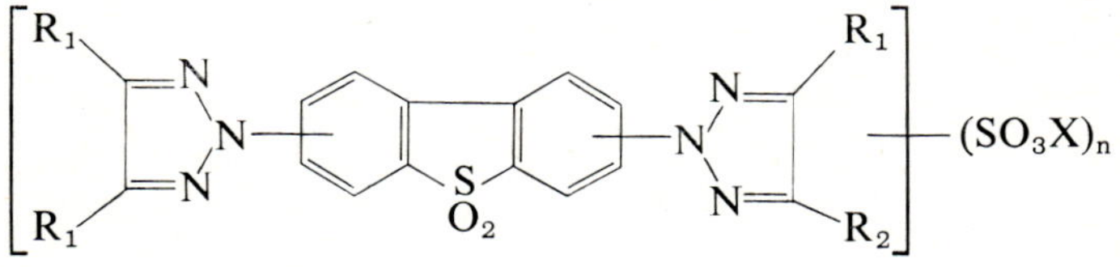

Fig. 2.46

11. Miscellaneous Structures Used or Claimed as FWAs

Oxacyanines (Fig. 2.47), familiar as photographic sensitizers, have been used for whitening textiles, particularly acrylics and polyamides[102]. The availability of many other FWAs, some of which show advantages over oxacyanines, has tended to restrict the use of the latter.

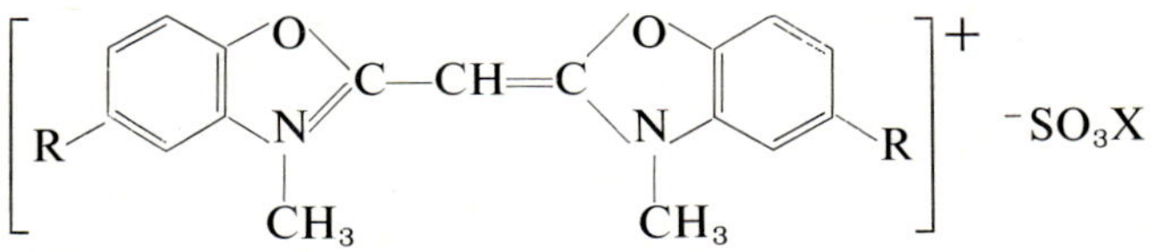

R = alkyl, alkoxy
X = aryl, —O—alkyl

Fig. 2.47

FWAs for use with polyesters are obtained by reaction (Friedel-Crafts) of pyrene with cyanuric chloride, followed by methoxylation of the condensed product[103] (Fig. 2.48). These products, however, have rather low light fastness.

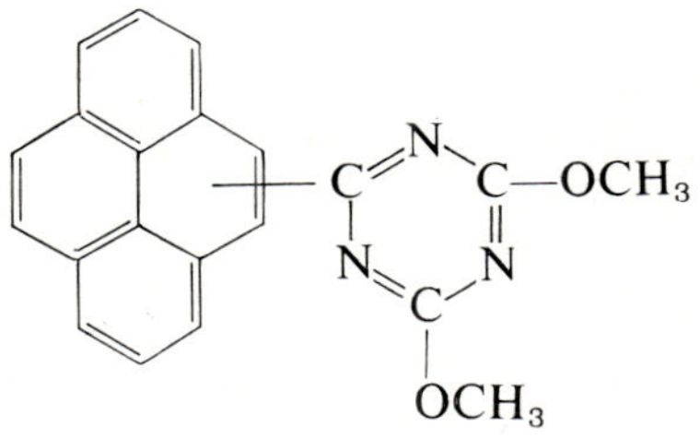

Fig. 2.48

Sulphonamide derivatives of the compound in Fig. 2.48 may be used as whiteners for nylon.

Triazinyl stilbenes(104) represented by the formula in Fig. 2.49 are prepared by reaction (Friedel-Crafts) of benzonitrile with stilbene 4:4′-dicarboxylic acid chloride.

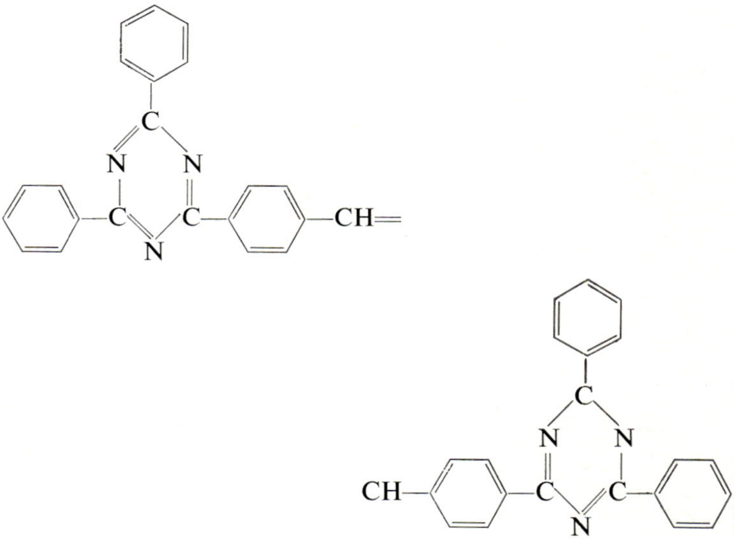

Fig. 2.49

The compounds which are very stable, have been recommended for use in polymer melts.

The fluorescent properties of many other structures have been investigated, but few products of commercial significance have been developed. Compounds studied include, for example, 2-arylbenzothiophene dioxide[105], diarylanthradipyridazone[106], acylated diaminopyrazine dicarboxylic acid[107], pyrrolines[108], dihydrocollidines[109], pyrazoloquinolines[110], substituted methines[111], triazolo-dihydroxypyrimidines[112], benzfuropyrazolines[113], pyrazolopyridines[114].

3

APPLICATIONS OF FWAs

FWAs are used to lighten and whiten many substrates. More than half the total production of FWAs is accounted for by household and laundry detergents. Textiles and paper use much of the remainder, with plastics and other applications accounting for only a very small percentage of the total production.

1. **Detergents**

Figures on the world production of soaps and detergents[115] provide a basis on which to estimate consumption of FWAs in this field. In North America, manufacturers tend to use a higher amount of FWA in heavy-duty soap and detergents, 0·3–0·5 per cent and more (on the weight of detergent) of full strength products; in Western Europe, the usage is of the order of 0·1–0·3 per cent and smaller quantities, 0·05–0·2 per cent, are used in other parts of the world. Of the hundreds of types of chemically-different compounds known and available, only about 25–30 different compounds are used to any significant extent by manufacturers of soaps and detergents. One of the main factors which determines the choice of FWAs for detergents is the composition of home laundry load; even in the U.S.A. (a country which uses a very high proportion of synthetic fibres) this contains more than 70 per cent cellulosics. For this reason, the bulk of the FWAs used in soaps and detergents are "cellulosic FWAs", which are mostly of *bis*-triazinylaminostilbene type. In countries where the home laundry load contains a substantial amount of materials made from synthetic fibres, FWAs effective on these fibres are also added in small quantities. Some of the FWAs which can be or are used in detergents are given in Table 3.1.

TABLE 3.1

I CELLULOSIC FWAS

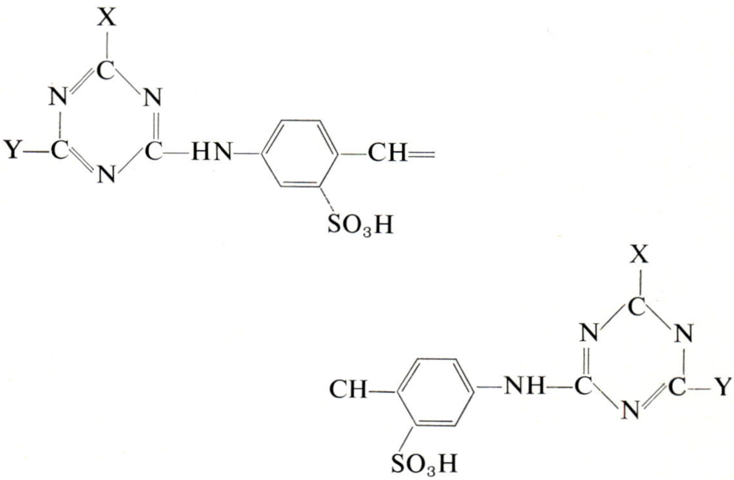

(a) X = —NH.PH; Y = —NHPh

(b) X = —N(CH$_2$CH$_2$)(CH$_2$CH$_2$)O; Y = —NHPh

$$X = -N\begin{matrix} CH_2CH_2 \\ \\ CH_2CH_2 \end{matrix}O; \quad Y = -NHPh$$

(c) X = alkanolamine, alkoxyalkylamine or alkylamine residue

e.g. $-N\begin{matrix} CH_2CH_2OH \\ \\ CH_2CH_2OH \end{matrix}$ or $-NHC_2H_5$ etc.

Y = —NHPh

(d) X = —OCH_3; Y = —NHPh

(continued)

Table 3.1 *(continued)*

II FWAS FOR POLYAMIDES

(a) *Pyrazolines*:

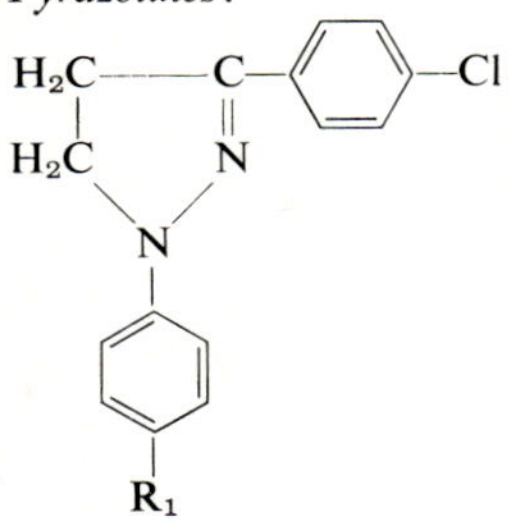

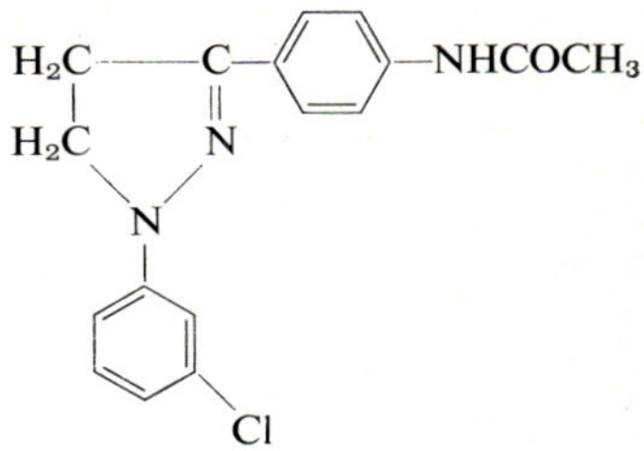

R_1 = $—SO_2NH_2$, or $—COOC_2H_5$

(b) *Naphthotriazoles*:

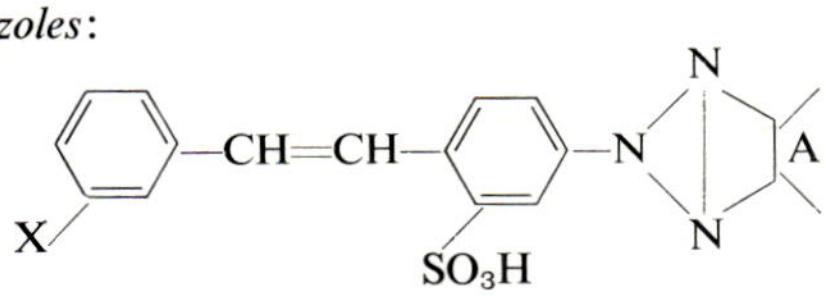

X = H or Cl

A = naphthalene or acenaphthene

(c) 1:2:3-*Triazoles and benzimidazoles*:

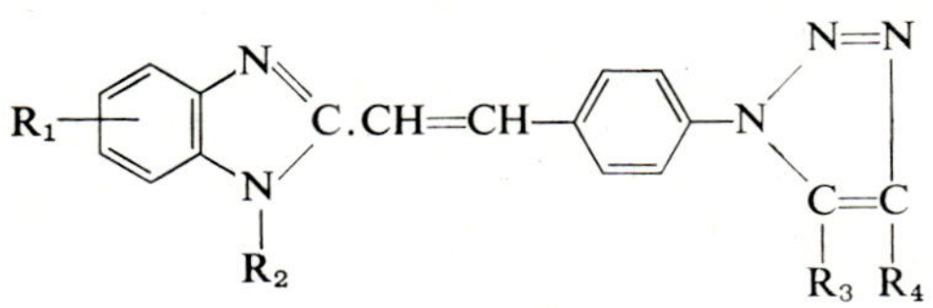

R_1, R_2, R_3, R_4 are H or alkyl or substituted alkyl

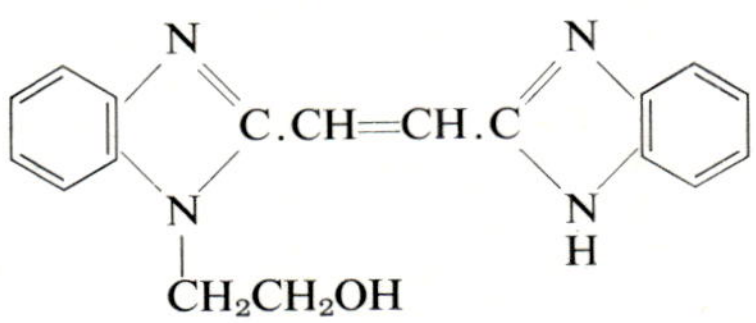

(continued)

Table 3.1 ***(continued)***

(d) *Coumarins and carbostyrils*:

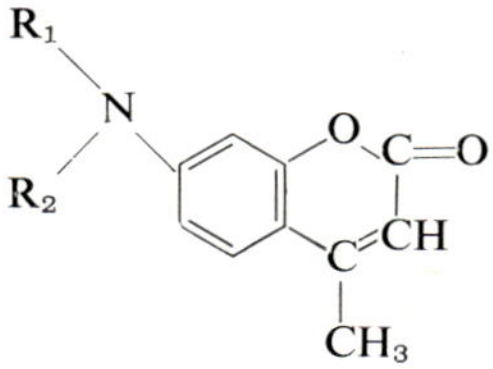

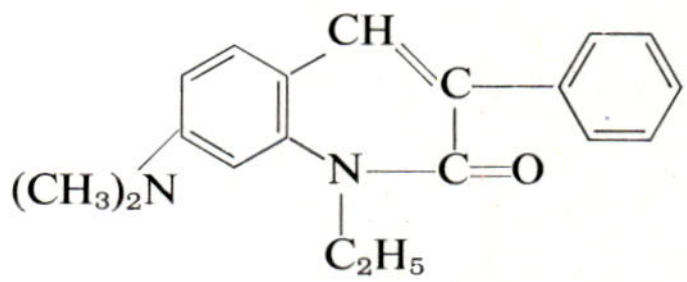

R_1, R_2 are lower alkyl

III FWAS FOR POLYESTERS, WHICH ARE ALSO EFFECTIVE ON POLYAMIDES

Polycyclic Oxazoles:

(a)

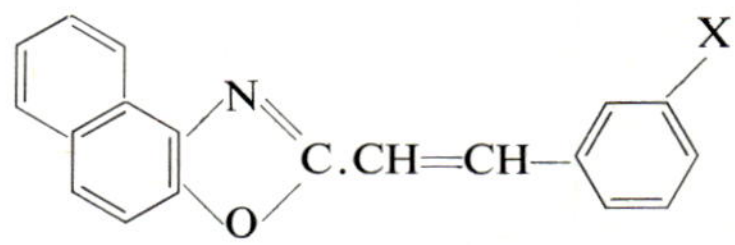

X = H or Cl

(b)

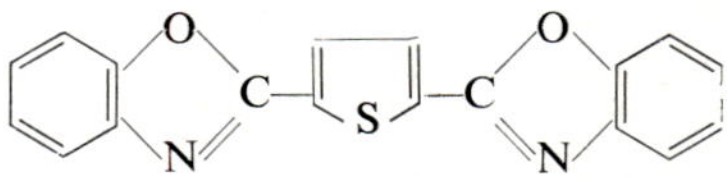

(c)

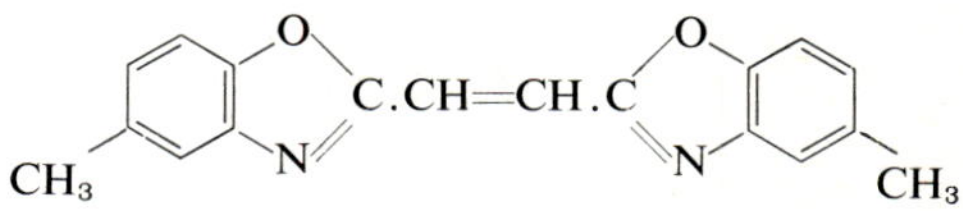

Coumarins

(d)

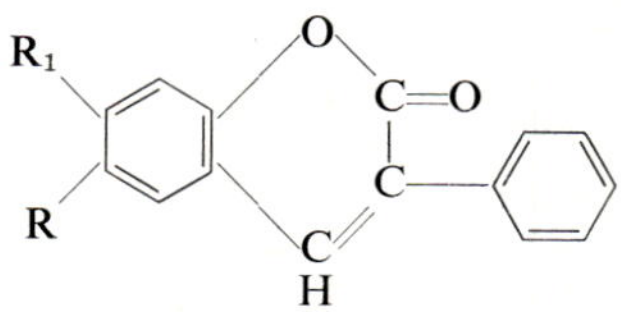

R = H or alkyl
R_1 = —SR_2 where R_2 = alkyl

Table 3.2 shows the composition of FWAs in some typical American detergents[116].

TABLE 3.2

Detergent	*Brightener type (full strength)*				
	Ib	Ic	IIb	IIIc	IIIa
	%	%	%	%	%
A	0·4	—	0·04	—	
B	0·2	—	0·1	—	
C	0·2	—	—	0·02	
D	0·3	—	0·04	0·02	
E*	—	0·2	0·1	—	0·03

* Cold water formulation.

N.B. In Europe, pyrazolines are used in many formulations in place of IIb and III where polyamide whitening is an important factor.

Factors affecting the choice of FWAs for detergent use have been discussed in detail by many authors[117–118]. Wash load composition is always an overriding consideration; other factors include:

1. Composition of the detergent.
2. Laundry wash conditions.
3. A variety of factors, such as hue of whiteness, light fastness, cost, toxicity, ease of incorporation, interaction of one FWA with another, whitening effect on the detergent powder, etc.

Composition of Detergent

Selection of an FWA depends to a large extent on the composition of the detergent. Compound Ia (Table 3.1) for example, is recommended as a FWA for cellulosics in anionic systems, but is less effective in non-ionic systems, Ib and Ic, on the other hand, are effective in both anionic and non-ionic systems.

The ability of the FWA to whiten a detergent depends on the surfactant system. In general, the whiteness of detergent powder obtained with commonly-used cellulosic FWAs is much better in the non-ionic than in the anionic-type detergent. Further, the exhaust of low-solubility FWA (type Ia) is lower in non-ionic than in anionic surfactants. A number of authors have investigated the effects described above(117–119).

Laundry Wash Conditions

Washing conditions vary greatly, depending on locality, and may range from handwashing in cold water to machine-washing with warm or boiling water. Moreover, the liquor-to-goods-and-detergent ratio varies from area to area. In the Table 3.3 Stensby(116) summarizes conditions used in different countries.

TABLE 3.3

Typical Laundering Conditions

Conditions	*U.S.A., Canada and Australia*	*Northern Europe*	*Mediter-ranean and Near East*	*Far East, Japan,* Latin America, etc.*
Detergent/cloth ratio	1:20–30	1:20–30	1:10–20	1:10–15
Cloth/liquor ratio	1:15	1:5–6	1:10–30	1:10–30
Wash procedure	Machine	Machine (heaters in machine)	Tub/hand machine	Tub-hand
Wash temperature	110–140°F	140–200°F	Mixed (cold or hot)	Cold (65–95°F)
Wash cycle	5–15 mins.	15–30 mins.	—	—
Hypochlorite	200 ppm	None	—	—

* Japan is moving into the hot machine-wash category.

Further complications are introduced in certain areas, such as North America, by the use of hypochlorite in the wash cycle. Most cotton FWAs are unstable to bleach, and in regions where the use of hypochlorite is common practice, household detergents will contain small amounts of cotton/polyamide FWAs of class IIb. In North America, FWA Ia was, until 1965, the most commonly-used product, but this has now been superseded by Ib. The change has taken place because (*a*) Ib has superior bleach stability, and (*b*) at high levels of incorporation it has superior build-up characteristics. FWA Ib is the preferred product under warm- or hot-water washing conditions, whereas types Ic and Id are probably more suitable for cold-water washing areas. The increasing use of resin-treated cotton and polyester/cotton blends presents the problem of whitening these fabrics by washing at 60°C or less; there is no really effective FWA available for these conditions.

As the amount of FWA used in detergents increases, the question of hue becomes more important. The work of Goldwasser(120) and other authors(121) suggests that, in the western hemisphere at least, white with a tinge of violet is commonly preferred. In many countries where a preponderance of white detergent powders is sold, the appearance of the product is an important factor in marketing detergents. Detergent whitening is related to (*a*) the structure of FWA used, (*b*) the composition of detergent, (*c*) the purity of the FWA, and (*d*) its physical form. The method of incorporation of FWA into a detergent may also play an important part.

With the introduction of detergents containing enzymes, there has been a growing tendency to favour FWAs which operate effectively both at low and at high temperatures.

2. **Natural Fibres and Regenerated Cellulose Fibres**

FWAs are essentially colourless fluorescent dyestuffs. The distribution of a coloured dyestuff within the fibre, however, is of less practical significance than the distribution of FWA within the fibre. FWA molecules which diffuse into the fibre contribute little to its whitening, and the FWA molecules should, as far as possible, attach themselves to the fibre surface. Also, agglomerations of FWA produce a "greening effect" which tends to diminish the whitening effect.

The method of application of a dyestuff is determined by its properties, but in the case of FWAs the method of application is often pre-determined by the finishing processes the fibre or the fabric must undergo. More often than not, FWAs should be capable of application under the conditions used in applying other auxiliaries.

2.1. *Cellulosic Textiles*

FWAs for application to cellulosic fibres behave like direct cotton dyes. Their uptake is influenced by a variety of factors, including the temperature and electrolyte content of the bath, liquor-to-goods ratio, etc. All FWAs used for cellulosic fibres are *bis*-(triazinylamino)stilbenes; for application purposes, they can be divided into three classes:

I. *Highly substantive*:

These products are affected only slightly by the electrolyte content or the temperature of the bath; they are therefore useful for exhaust processes.
Chemical type: for example, Table 2.1. Types 5 and 9.

II. *Medium (low) substantive*:

These are strongly influenced by the electrolyte content of the bath, but less so by its temperature; they are therefore useful for padding processes.
Chemical type: for example Table 2.1. Type 10.

III. *Low (medium) substantive*:

The uptake of these FWAs decreases with increase of temperature, and is less affected by the electrolyte content of the bath. They have particularly good levelling properties, and are therefore useful for pack machines.

Chemical type: for example Table 2.1. Types 3 or 8.

Note: Not all FWAs fit the above description; some of them, for example, show properties of both groups II and III.

Mantz and Brüggemann[122] have summarized the methods of application of various types of FWA (see Table 3.4).

TABLE 3.4

Methods of Application of FWAs

Dyeing machine	*Method of application*	*Recommended FWA type*	*Tips for application*
Winch	Cold liquors	II, (I)	
	Hot liquors	I, (II)	
Rope dyeing machine	Cold and hot liquors	II, I	
Jigger	Cold liquors	II, (I)	FWA and salt to be added at the two ends
	Hot liquors	I, (II)	FWA added at four ends
Cheese and beam dyeing machine	Hot liquors, e.g. peroxide bleach	I, (II)	FWA added in the bath
Pack dyeing machine	Hot liquors, e.g. peroxide	III, (I)	Liquor (ready with FWA) forced from underneath
Continuous winch and wash batteries	Cold to warm liquor	I, II	About one-fifth of the FWA required for 1 ton material added before—the rest added later continuously
Pad-mangle	From aqueous liquor or from liquors containing finishing agents	II	Check the compatibility of the FWA with the finishing agents used

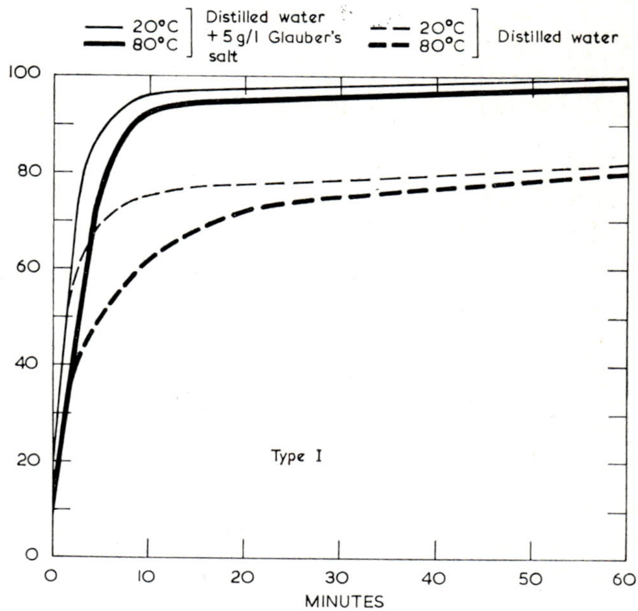
20°C
80°C
Distilled water + 5 g/l Glauber's salt
20°C
80°C
Distilled water
100
80
60
40
20
0
Type I
0
10
20
30
40
50
60
MINUTES

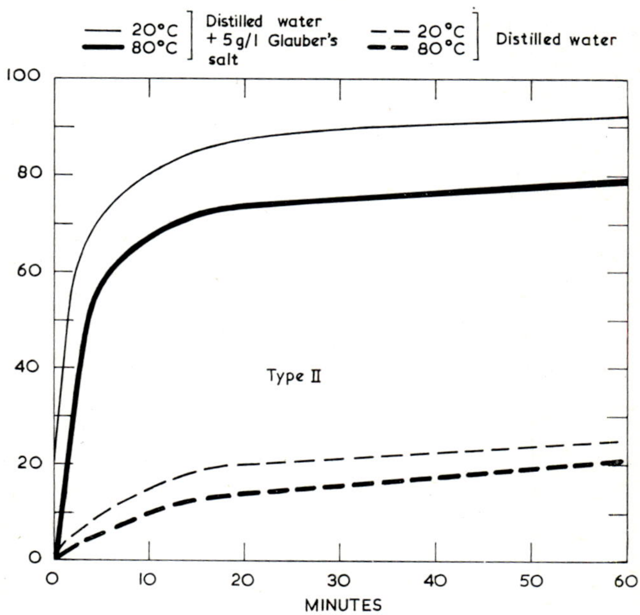
20°C
80°C
Distilled water + 5 g/l Glauber's salt
20°C
80°C
Distilled water
100
80
60
40
20
0
Type II
0
10
20
30
40
50
60
MINUTES

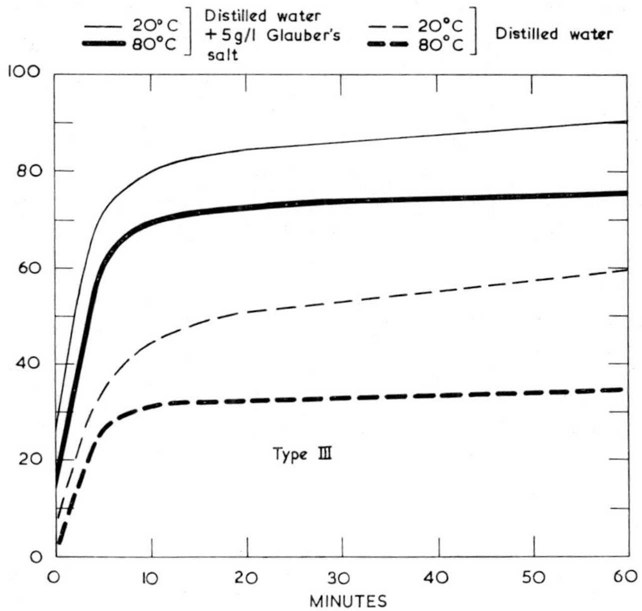

Fig. 3.1. (above and opposite page)

(After Mantz and Bruggemann[122]*, SVF Fachorgan (1964))*

To obtain a very high degree of whiteness, which the consumer demands today (particularly for shirtings and knitted goods), an extremely well-bleached ground is necessary. This is prepared usually by the following sequence of operations:

Scour—hypochlorite bleaching—peroxide bleaching or scour and chlorite bleaching.

Most FWAs used by the textile (cellulosic) finishing industry are unstable in the presence of hypochlorite or chlorite, but are stable to peroxide; they can be added to peroxide bleach baths. When treatment with an FWA follows hypochlorite or chlorite bleaching, it is customary and necessary to give the fabric or goods an "antichlor" treatment with bisulphite, or an acid sour before treatment with FWA. This pre-treatment can be dispensed with when peroxide bleaching follows the hypochlorite treatment; the FWA can then be added to the peroxide bath. The sequence of treatments indicated above lends itself to semi-continuous or continuous operations. The most brilliant effects in this type of treatment are obtained by the use of a FWA of type I.

2.2. *Cellulosic Textiles Treated with Finishing Agent*

With the advent of "easy care" textiles, more and more cotton is being given crease-resistant finishes. This can create problems, as application of an FWA from a resin bath may not produce as good a white as that obtained when the FWA is applied by a normal application process. This is partly due to the fact that many FWAs are to some degree acid-sensitive. The FWAs used in such finishes normally should be applicable by a padding process; they should have a high solubility and consequently not too high a fibre substantivity. They should, however, have affinity for the fibre under weakly-acid conditions, and they should not be adversely affected by heating to temperatures used for curing the resins. When FWA is added directly to the resin bath, adequate stability to resins like N-methylol compounds and the usual catalysts is desirable. The whitening of cotton which is to be resin treated may be carried out either by a one-bath or a two-bath process. In one-bath processes, the FWA is added directly to the resin bath. In the two-bath process, the main whitening is carried out after or during bleaching, and a small amount of FWA is added to the resin bath. The first bleaching and whitening in the two-bath process can be carried out in the normal manner. Optimum results are obtained if the goods are dried after such treatment and before resin finishing.

The one-bath process, as the name suggests, is very simple in practice. Whitening and resin-finishing are carried out in the same bath (after chemical bleaching). This process needs greater care in the choice and quantity of FWA used, and the possibility of interaction between FWA and resin must be considered.

Choice of the FWA in resin-finishing operations depends to a great extent upon the catalyst employed, which may for example include ammonium sulphate, ammonium chloride, diammonium and mono-ammonium phosphate, zinc nitrate and zinc chloride, magnesium chloride.

FWAs used for this purpose normally yield a stable solution in the presence of ammonium salts. They may produce a greyish colour when the curing is carried out at higher temperatures and for a prolonged period. The use of zinc nitrate causes yellowing when the goods come in contact with light, and in some cases the light fastness is reduced. This can be overcome by washing the material after resin treatment. Zinc nitrate also precipitates many FWAs, and it is therefore used mostly in two-bath processes which conclude with a thorough soaping. Zinc chloride affects the stability of many FWAs; it is particularly effective for two-bath processes, as it does not diminish the whitening effect, even at high temperatures. Magnesium chloride is the best catalyst for one-bath processes, providing a very stable solution with FWAs; the whitening effect is stable to high temperatures.

Other finishing agents often used include softening agents and agents which improve the handle of cellulosic material. Cellulosic FWAs may be adversely affected by cationic auxiliary and emulsifying agents, but are quite compatible with anionic or non-ionic agents. The finishing agents should have no self-colour, otherwise they affect the whiteness of the finished material. Finishing agents commonly used for improving the handle of cellulosic materials are based on starch, cellulose, polyvinyl alcohol or acetate, polyacrylic esters, polyethylene, etc. Cellulosic FWAs have affinity for the starch and cellulose-based products, to which they impart some whitening effect. Agents based on dispersions of polymeric material (and not containing cationic emulsifying agents) have only slight affinity for the FWAs.

2.3. *Wool*

Wool yellows on exposure to sunlight. Opinions differ regarding the effect of FWA on the yellowing of wool; some authors believe that FWAs accelerate the effect[(123)], others suggest that, under certain conditions of application, FWAs do not have such a drastic effect[(124)]. Wool is normally subjected to a peroxide (oxidative) bleach or a reductive bleach—hydrosulphite or stabilized hydrosulphite or a combination of both. An oxidative bleach followed by a reductive bleach with a stabilized hydrosulphite type is commonly preferred. FWAs are generally applied to wool with the reductive bleaches under acid or neutral conditions depending on the nature of the particular FWA used. Chemical types used for the whitening of wool fibre include:

1. *bis*-(triazinylamino)stilbene type
2. Pyrazolines
3. Amino coumarins

Amino coumarins have very poor lightfastness.

The sequence of operations used in bleaching and whitening of wool is as follows:

(*a*) washing with 1–2 g/1 detergent and 1–2 ml/1 ammonia at 40°C for about ½ hr,
(*b*) peroxide bleaching (10–20 ml/1 H_2O_2; pH 8·5 at 45–50°C),
(*c*) reductive bleaching and whitening (3–5 g/1 stabilized hydrosulphite and 0·5–1·5 per cent FWA at 45–53°C for 2–3 hr).

During reductive bleaching, a little acid (1–2 per cent acetic or formic acid) is added ½–1 hr after the beginning of reduction.

3. **Synthetic Fibres**

Production and consumption of man-made fibres is increasing rapidly. In 1969, the world production of synthetic fibres surpassed, for the first time, the production of man-made fibres based on cellulose. During that year more than 70 per cent of the total production of FWAs was of the type which can be applied to cellulosic fibres, but the demand for FWAs for whitening synthetic fibres is increasing rapidly.

3.1. *Polyamides*

Polyamides, at the beginning of the 1970s, remain the most important wholly-synthetic fibre group. The demand for FWAs used in whitening polyamides is greater than that for whitening any other type of synthetic fibre.

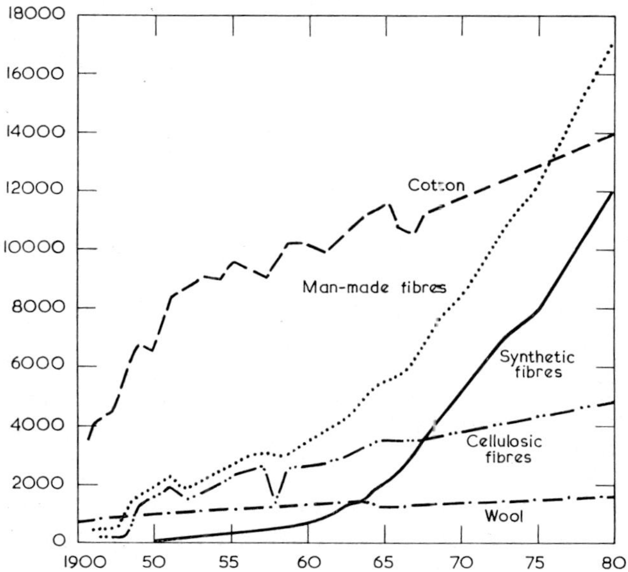

Fig. 3.2. World production of cotton, wool and man-made fibres, in 1000 tons

(*After Chemische Industrie, 1969*, 21, *823*)

TABLE 3.5

World Consumption of Textile Fibres
(1000 metric tons)[125]

	1966	*1967*	*1968*	*Percentage Share 1968*	*Percentage Change 1968/67*
Cotton	11,120	11,320	11,450	57	1
Wool	1,540	1,463	1,524	7	4
Manmade: Cellulosics	3,338	3,307	3,500	17	6
Manmade: Synthetics	2,479	2,862	3,750	19	31
Total	18,477	18,952	20,224	100	

TABLE 3.6

World Production of Synthetic Fibres
(1000 metric tons)[125]

	1966 Percentage	*1967 Percentage*	*1968 Percentage*	*Change 1968/7 in Percentage*
Polyamide	1,214 (49)	1,314 (46)	1,615 (43)	+23
Polyester	589 (24)	753 (26)	1,085 (29)	+44
Acrylics	457 (18)	541 (19)	728 (19)	+35
Other	219 (9)	254 (9)	322 (9)	+27

FWAs used commercially for polyamides can be divided into two types.

1. *Acid Type*: FWAs in this category are of the *bis*-(triazinylamino)-stilbene, pyrazoline, 1:2:3-triazole, or *bis*-styryl-biphenyl types.
2. *Disperse Type*: FWAs in this category are mostly pyrazolines or derivatives of amino-coumarins.

Bis-(triazinylamino)-stilbenes are, from the point of view of the textile finisher, the most important FWAs for polyamides. They are inexpensive, but give a very brilliant white effect. Pyrazolines, 1:2:3-triazoles and *bis*-styrylbiphenyl are used to a limited extent.

The disperse types are used mainly for incorporation into detergents, but they may also be used by the textile finisher. These FWAs are, in general, more expensive than the acid types.

Bis(triazinylamino)stilbenes are applied to cotton as direct dyes, but they are applied to polyamides as acid dyes; they are held by the fibre as a result of salt formation.

Products of higher affinity (such as Table 2.1, type 5) can be applied under neutral or weakly acid conditions; reducing agents such as hydrosulphite may be used, the hydrosulphite providing a solution of adequate acidity.

Products of lower affinity (such as Table 2.1, type 1) require conditions of higher acidity (pH around 4).

The type of nylon to be whitened influences the degree of acidity required. Intense whites are more easily obtained on nylon 6 than on nylon 6·6; high-temperature conditions may be needed to obtain high whites with nylon 6·6. FWAs of the disperse type are preferred for nylon 11.

Bis(triazinylamino)stilbenes and pyrazolines are attacked by chlorine-containing bleaches. Nylon is commonly bleached with reductive bleaches, e.g. hydrosulphite; most acid types of FWA can be added to such baths. If a chlorite bleach is essential to obtain a good ground, the FWA may be added either to the "antichlor" (normally bisulphite) bath or, better still, after the "antichlor" bath.

FWAs may be applied to polyamide fibres by the following methods:

1. Hydrosetting process (H-T equipment).
2. Finishing on the jigger or winch beck with thermo-setting.
3. Continuous or semi-continuous processes.

Hydrosetting Process. The following sequence of operations is used for whitening polyamide fabric by this method.

Pre-washing; pre-stretching and winding; autoclave treatment; drying.

The preliminary scouring can be omitted in the case of lightly-soiled goods. With knitted goods, it is advisable to pre-set or stabilize the lapping at temperatures close to normal heatsetting temperature of the polyamides in question. Whenever pre-washing is omitted, the goods must at least be steamed at the feed end of the stenter. If the goods are clean, therefore, washing, bleaching, whitening and hydrosetting are combined in a single operation.

The following are two typical recipes:

1.	FWA (high affinity)	1–3 per cent
	Hydrosulphite	3 g/litre
	Surface active agent (for synthetic fibres)	0·5 g/litre
2.	FWA (low affinity)	1–3 per cent
	Hydrosulphite	3 g/litre
	Surface active agent	0·5 g/litre
	Acetic acid	0·5 g/litre (or pH = 4 adjusted with acetic acid)

Hydrosetting is carried out for 15–20 minutes at 120–130°C (120°C for nylon 6, 130°C for nylon 6·6). The treatment liquor is cooled to 80–85°C and the material squeezed (rinsed) and dried. Before final drying, some materials (e.g. shirting, etc.) are treated with auxiliaries (such as anti-statics, softeners, etc.) on the padding mangle.

Finishing on the Jigger and Winchbeck. Recipes used in these processes are similar to those set out above, except that high-affinity products only are recommended. Roller vats, which allow a longer residence time (Fig. 3.3), yield more satisfactory results.

Continuous and Semi-continuous Methods. Selected high-affinity acid-type (*bis*(triazinylamino)stilbenes and pyra-

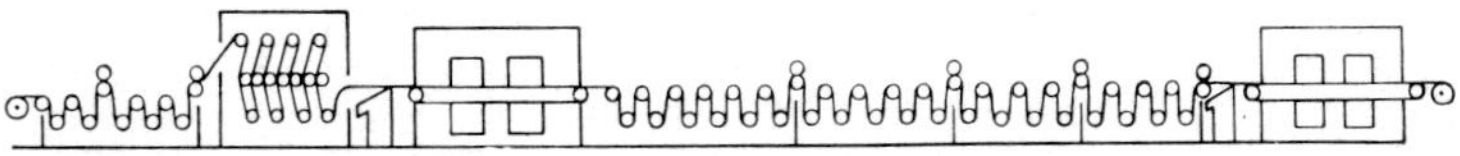

Fig. 3.3. Finishing on the jigger and winchbeck

(After SVF Fachorgan 1964, 19, (6), *437)*

zolines) and dispersed-type FWAs can be applied to polyamides by continuous or semi-continuous processes such as (*a*) pad-roll process, or (*b*) pad-steam process.

Pad-roll Process. Padding is carried out at 50–70°C, with an expression of 60–70 per cent. Full development of the FWA is ensured by keeping the goods in a developing chamber for 1–2 hr at 90°C. Ideally, acid conditions are maintained during dyeing and development, e.g. by adding ammonium chloride. The goods are washed in hot water before being passed through a stenter.

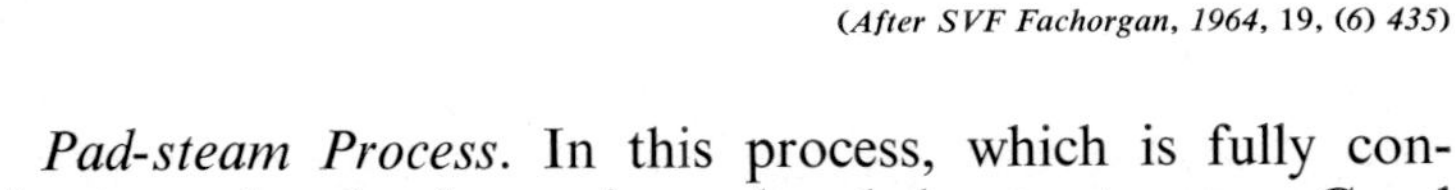

Fig. 3.4. Pad-roll process

(After SVF Fachorgan, 1964, 19, (6) *435)*

Pad-steam Process. In this process, which is fully continuous, the developer is replaced by a steamer. Good results are obtained by steaming for about 3 minutes at a steam temperature of 105°C, or for shorter periods at a highers team temperature. Acid-releasing agents, e.g. ammonium chloride, yield higher whites.

Thermosol Process. Certain acid type FWAs (pyrazolines and 1:2:3-triazoles) can be applied to polyamides by the thermosol process. The polyamide is padded at pH 4 with a solution of the FWA, dried and heat treated for 30 sec on a stenter at 190°C.

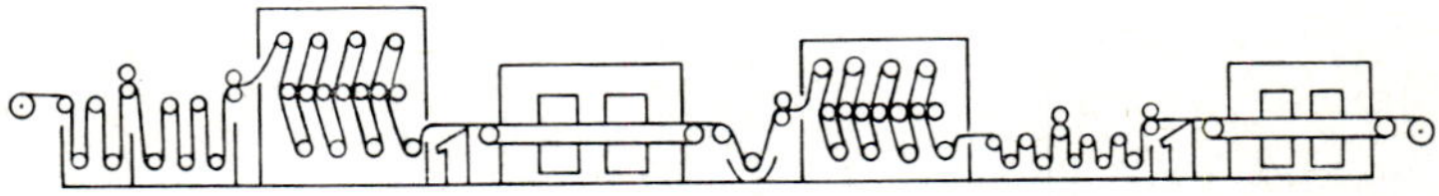

Fig. 3.5. Pad-steam process

(After SVF Fachorgan, 1964, 19, *(6), 436)*

3.2. *Polyamide/Wool Blends*

These are generally whitened from reductive bleaching baths, using acid type FWAs under weakly acid or neutral conditions.

3.3. *Polyamide/Cotton and Polyamide/Viscose Blends*

Blends of polyamide with viscose or cotton may be whitened with FWAs having high affinity for both the fibres; they can be applied from reductive bleaching baths.

3.4. *Polyesters*

Polyester materials are whitened by selected disperse types of FWA. Recommended types are listed in Table 3.7.

Polyester materials may be whitened by exhaust processes at the boil (97°C) or under high temperature conditions, with or without a carrier, depending on the nature of the product. In most cases, where high-temperature conditions are not used, better results are obtained with the aid of carriers. Where high-temperature conditions are used, however, carriers are not generally needed.

Under high-temperature conditions, the material is normally introduced into the bath at 50–70°C; the temperature is raised to 120–130°C within $\frac{1}{2}$ hr and held at that temperature for a further $\frac{1}{2}$ hr. When the fabric is soiled, or the starting background is inadequate, a chlorite bleach may be incorporated with the whitening. In many cases, chlorite bleaching gives a yellower background to the polyester material. The effect may be influenced by the catalyst used in making the polymer.

TABLE 3.7

I. *Polycyclic oxazoles*

(a)

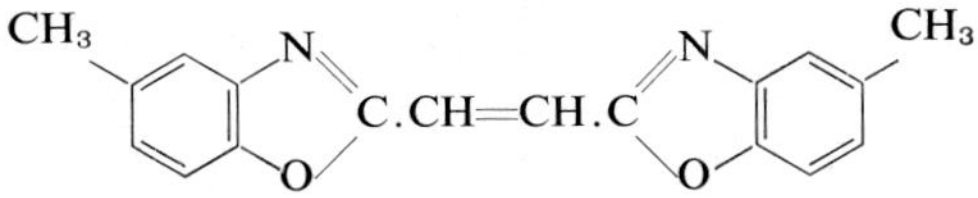

(b)

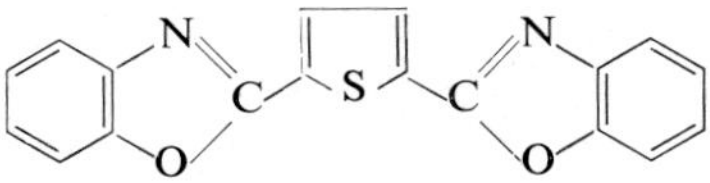

(c)

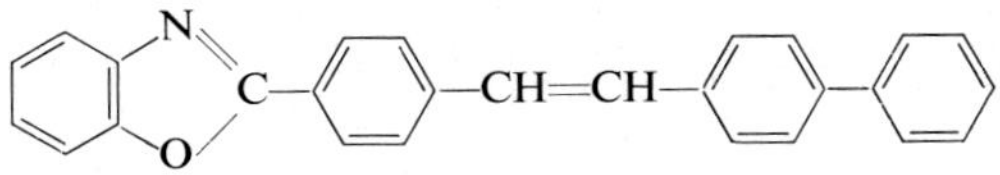

(d)

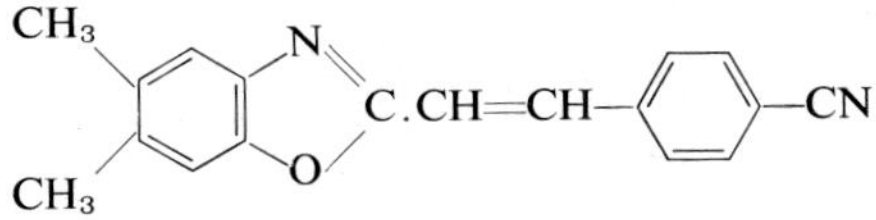

(e)

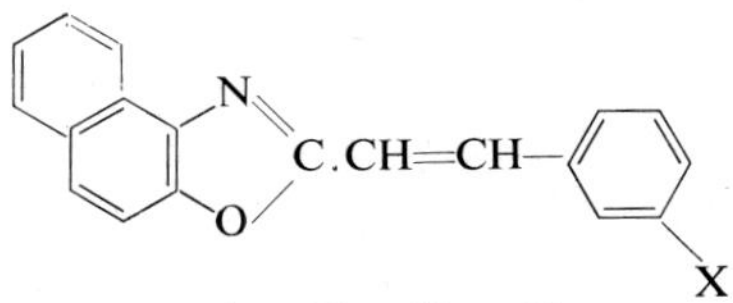

where X = H or Cl

II. *Coumarins*

(a)

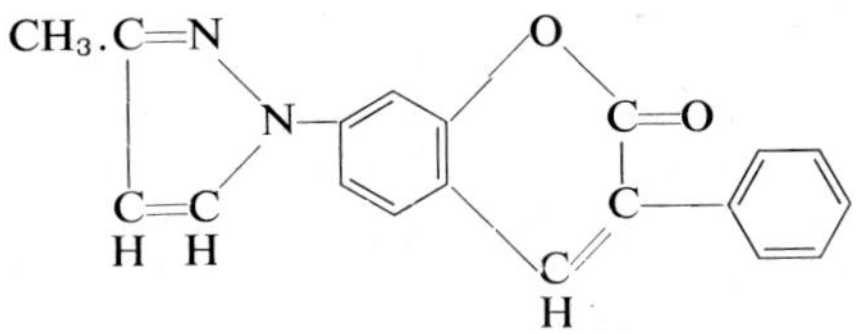

(b)

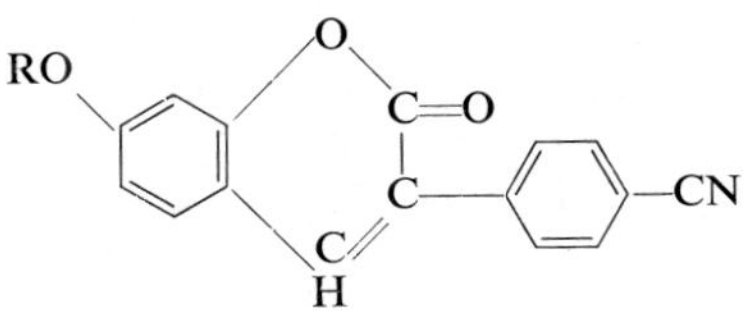

(continued)

Table 3.7 *(continued)*

(c)

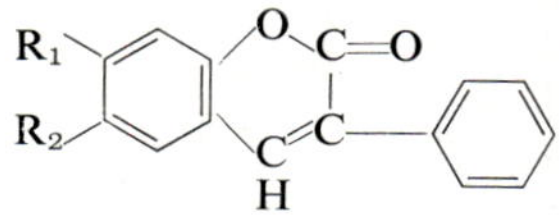

$R_1 = -SR$; $R_2 = H$ or alkyl; $R = $ alkyl

III. *Aryltriazoles*:

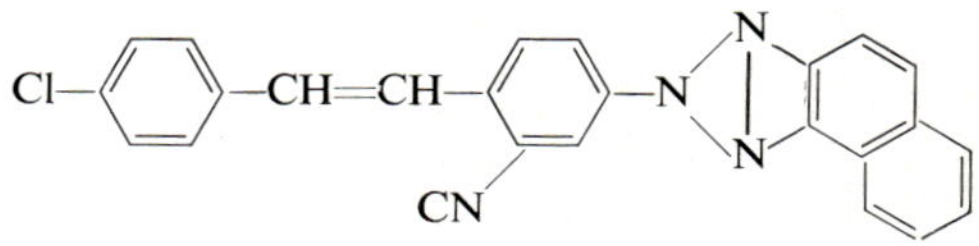

IV. *Styryl stilbenes*:

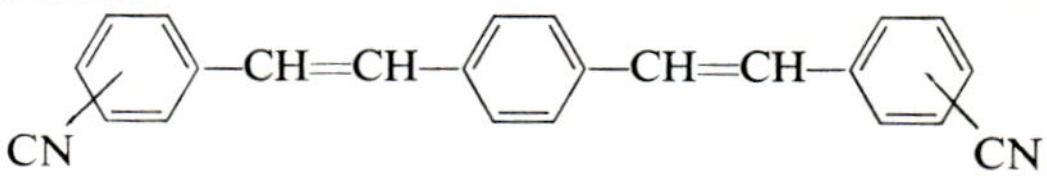

V. *Naphthalimides*:

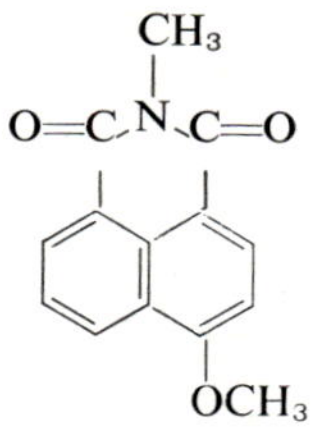

VI. *Triazinyl pyrene*:

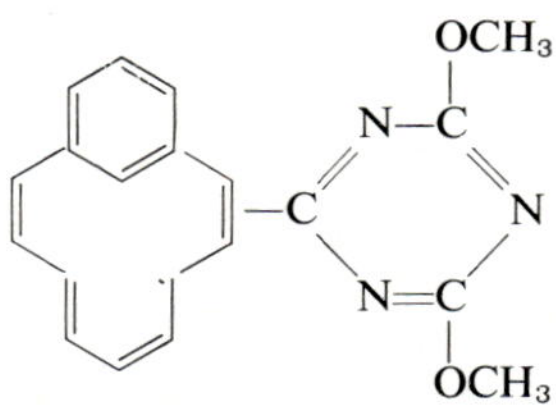

N.B. IIa and III are sold as a mixture.

The *thermofixation* or *pad-bake* process is an important method of applying FWAs to polyester fabric. The fabric is padded with a dilute suspension of the FWA, dried in an open stenter and then heated for ½ to 2 minutes at 170–220°C. This process makes possible a higher throughput of material with superior dimensional stability.

Properties of some of the compounds mentioned above are summarized in Table 3.8.

TABLE 3.8

Degree of Whitening

	Application by Exhaust		*Application by Pad-bake*		*Chlorite stability*	*Light fastness*
Product	*at the boil*	*H.T.*	<200°C	>200°C		
Ia	High	High	High	Low	Good	>6
Ib	High	High	High	Medium	Good	>6
Ic	Low	Low	Low-medium	High	Fair	6
Id	High	Medium	High	Medium	Good	>6
Ie	High	High	High	Medium	Fair	4–5
IIa + III	Low	High	Low	High	Good	>6
IIb	Medium	High	High	High	Good	>6
IIc	High	High	High	High	Poor	5–6
IV	Low	High	Low	High	Good	5
V	Medium	Medium	Medium	Low	Good	6
VI	High	High	High	High	Poor	4

In many countries, particularly in North America, increasing amounts of polyester materials are being treated at the melt-spinning stage. Many types of FWA are used, those shown in Fig. 3.6 being examples. Some of the compounds in Table 3.7 are also suitable for this purpose.

3.5. *Polyester/Cotton Blends*

Fabric made of polyester/cotton blends (50:50 or 65:35) is rapidly gaining popularity in many countries, much of it being used for white shirtings and sheets. Fabrics of this type

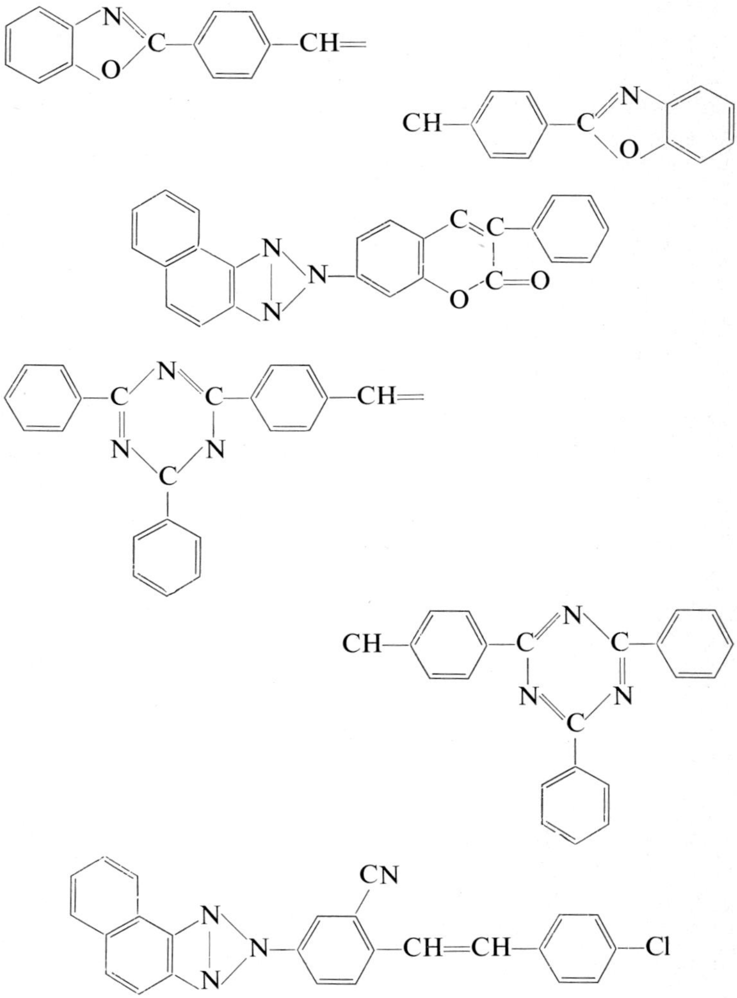

Fig. 3.6

are commonly whitened by using a combination of an FWA for polyester and an FWA for cotton. The FWAs may be applied during one of the bleaching processes or during resin finishing. Many polyester FWAs may be applied during (in conjunction with) the chlorite bleaching, whereas the cellulosic FWA is generally added to the oxidative or reductive bleaches. Both FWAs may be added to the resin bath; in this case, it is necessary to wash the fabric after resinating to remove deposits of precipitated FWA from the surface of the cloth. A typical sequence of operations is as follows:

Washing — mercerizing — bleaching — washing — drying —"resination" (FWA added)—curing—washing—drying.

The cellulosic FWA may also be applied with a reductive bleach, e.g. hydrosulphite, during the final washing in roller vats before drying.

3.6. *Acrylic Fibres*

Acrylic fibres are obtained by the polymerization of acrylonitrile and minor amounts of other vinyl compounds. The properties and behaviour, e.g. to dyeing, of an acrylic fibre depend upon its chemical constitution, and individual fibres vary considerably in these respects. The techniques used for whitening acrylic fibres may differ widely, specific FWAs and methods of application being selected to suit individual fibres.

Modacrylic fibres, containing smaller proportions of acrylonitrile, must be treated individually in a similar way.

About 4·6 per cent of the world production (1970) of acrylic fibres is treated with FWAs, which may be grouped into two classes:

1. *Disperse type*, including pyrazolines, naphthalimides and aminocoumarins.
2. *Basic and/or Cationic type*, including pyrazolines, naphthalimides, oxacyanines, oxazoles and aminocoumarins.

6 20pp.

As in the case of other fibres, the process used in whitening acrylic fibres depends upon their self-colour. Some acrylics have an excellent white self-colour (background), and they can be whitened directly; others with poorer backgrounds require bleaching, generally with sodium chlorite. (Chlorite bleaching is not recommended for Courtelle.)

Some of the FWAs on the market are chlorite-stable, and others are not. Chlorite-stable products may be applied from a bleach-bath; chlorite-unstable products are used when chlorite-bleaching is unnecessary, or following a chlorite bleach. Most FWAs are applied to acrylic fibres under acid conditions; oxalic, acetic and formic acids (1 g/l) are often recommended. Acrylic fibres are commonly treated by exhaust methods at the boil (>90°C). Liquor ratios of 1:40 are used, the material being introduced into the bath at about 70°C; the liquor is brought to the boil in ½ hr and held at that temperature for a further ½ hr. The bath is finally cooled and the material washed or rinsed. (In the case of Courtelle, it is recommended that sodium bisulphite (0·25 g/l) and sodium acetate (1 g/l) should be added to the bath in addition to the acid).

3.7. *Cellulose Acetates*

Cellulose acetate and triacetate fibres are generally whitened with selected disperse-type FWAs, many of which can be used for other synthetic fibres. The FWAs are commonly of the following types: pyrazolines, amino coumarins, benzoxazoles and naphthalimides. Whitening of triacetates is normally carried out by an exhaust process, but some FWAs may also be applied by the "thermosol" process at 200°C for 30 seconds.

3.8. *Polyvinyl Chloride and Polyolefin Fibres*

FWAs used for polyesters are often applicable to these fibres under usual exhaust conditions.

4. Paper

FWAs are used extensively in the paper industry to improve the whiteness of paper. The types of FWA used are mainly those with an affinity for cellulosic fibres—viz: *bis*-(triazinylamino)stilbenes.

FWAs can be introduced during one or more of the following stages of paper manufacture:

1. Pulp manufacture.
2. Paper making: (*a*) by adding to the beater,
   (*b*) by adding to the size press,
   (*c*) by coating on the machine,
   (*d*) by spraying or water-box treatment at the calendar.
3. Paper converting or coating operations.

FWAs used for whitening paper have affinity for the groundwood as well as for any other pulp. Under suitable conditions, exhaustion of more than 90 per cent can be obtained on bleached or unbleached pulp—sulphite, kraft or groundwood. FWAs can add only the final touch in improving the whiteness of the pulp; unbleached pulp cannot be whitened satisfactorily. This lack of effect on groundwood or unbleached pulp is caused by substances of the lignin type which act as UV-absorbers.

Factors which influence the efficiency of FWAs when added in the beater include purity and pH of water, alum size, fillers, wet-strength resins, dwell time, temperature, order of addition, concentration, etc.

In certain areas, water contains a significant concentration of iron ions. This has a deleterious effect on the whitening efficiency of FWAs. In peaty districts, the water may contain abnormal amounts of humic acids which diminish the whitening effect by absorbing UV radiation.

Opinions differ regarding the effect of hardness of water on the whitening efficiency of FWAs. Hardness can increase the whitening effect of some FWAs, but in other cases the whitening effect may be diminished. Hardness of water reduces the solubility of many FWAs, which should be dissolved separately in soft water. Many manufacturers sell liquid products (solutions of FWA), which overcome this problem.

The effectiveness of FWAs varies with pH, the degree of variation being different from product to product. Many FWAs which are highly effective at pH 6–8 may be relatively ineffective at pH 4–4·5.

FWAs for whitening paper should possess good fluorescence even in an acid medium. FWAs are commonly marketed in the form of an alkali metal salt. If, under the conditions of paper manufacture, the alkali salt is converted to free acid before it is able to affect the fibre, the FWA will become ineffective. In most paper mills, the pulp is entered into the hydropulper and beaten with back water, usually at pH 5 to 5·5. The effectiveness of the FWA is dependent largely on this initial pH, rather than the final pH. Unsized paper is always whiter than sized paper. Rosin size, being a UV-absorber, reduces the effectiveness of FWAs; careful selection of rosin size is therefore important. Some FWAs may flocculate if alum is employed in large quantities; they should be added before the alum. In continuous coloration processes, the FWA is added last, shortly before sheet-making; usually there is a slight diminution of whitening.

Fillers are often used in white papers. China clay or kaolin is commonly used; others include titanium dioxide, albafixe and talcum. In the production of white paper it is preferable to avoid using fillers with a yellowish self-colour. Each FWA will react to clay in its own particular way. Titanium dioxide absorbs in the UV range, the Rutile form absorbing more strongly than the Anastase form. Rutile is therefore less suitable as a filler when an FWA is used. Calcium and magnesium silicate fillers have the least adverse effect on FWAs.

Cation-active resins are often used as retention aids, and to improve the wet strength of paper. Addition of such resins causes a diminution of whitening effect, urea-formaldehyde having less effect than melamine-formaldehyde types.

Different products need different times for complete exhaustion. Some FWAs exhaust less readily at lower than at higher temperatures, but most products now on the market show negligible differences between 40 and 100°F.

At lower concentrations, the degree of whitening normally increases steeply with the increase in concentration of the FWA (Fig. 3.7). Above a certain concentration, however, the whiteness may remain unchanged or even diminish as concentration increases. The level of concentration at which this happens depends upon the product and on the conditions used.

Application of FWAs on the Surface of Paper

FWAs may be applied on the surface of paper, a technique which has become commercially important.

Paper is commonly coated to improve its printing properties. Coating may mask the whiteness of the paper itself, and it is thus more effective to apply FWA to the surface coat. This may be achieved by (*a*) surface sizing, or (*b*) surface coating.

Surface sizing is carried out usually in the size-press. The paper is impregnated or sprayed, and then pressed with liquor containing water-soluble sizing compositions, such as glue, starch or carboxymethyl cellulose. FWAs are added to the impregnation liquors. The operation may be carried out outside the paper machine, e.g. in the calender. FWAs used for this purpose should have an affinity for the sizing material; CMC and starch are particularly suitable. Factors affecting whitening efficiency in this type of application include temperature and viscosity of impregnation liquor, immersion time, pressure exerted by the squeezing rollers and viscosity of the liquor. The pH of the dip bath is important; it should not be allowed to fall below 5–6, and may be adjusted by the addition of ammonia.

The whiteness of the paper before sizing is important, as

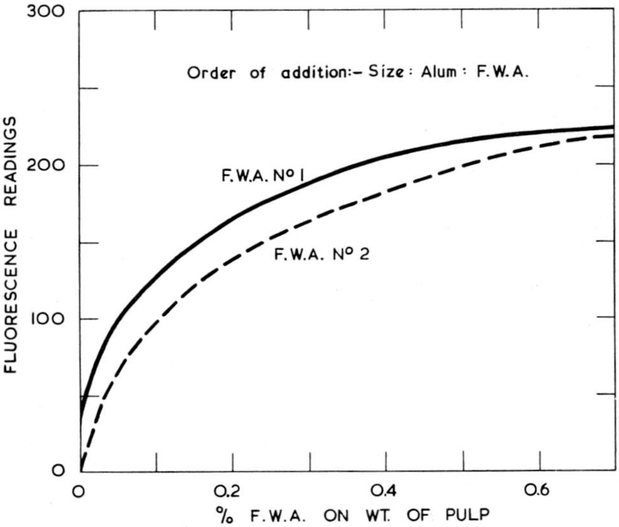

Fig. 3.7

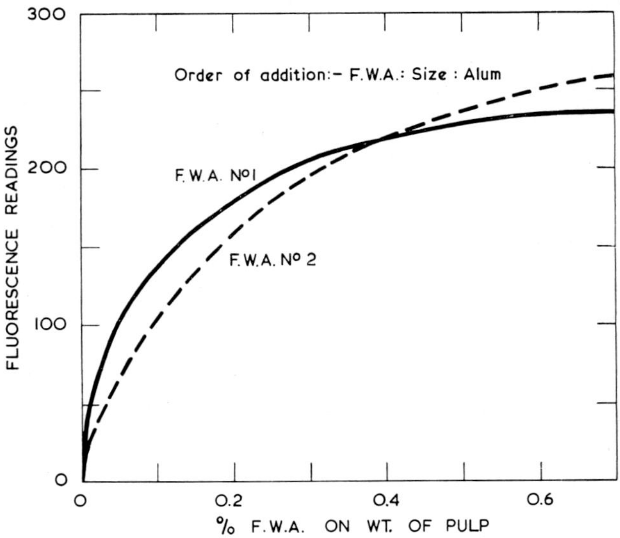

Fig. 3.8. At lower concentrations the degree of whitening normally increases steeply with the increase in concentration of the FWA

the coating obtained by surface sizing is transparent. Shading dyestuffs (blue or violet) are sometimes added to the impregnation liquor.

Surface coating is used for producing high-grade printing papers. The coating consists usually of white pigments mixed with binding agents and water as follows:

White Pigments	*Binding Agents*
Clay, Satin white, barium sulphate, titanium dioxide, calcium carbonate, etc.	Usually a mixture of natural and synthetic binders. Natural binders are commonly casein and starch. Synthetic binders are dispersions of polymeric material based on acrylic acid esters and styrene-butadiene polymers.

Binding agents have a major influence on the whitening of coating pastes. Starch and casein differ in their reaction to individual FWAs. Casein is yellower than starch, but FWA coatings containing casein are whiter than those containing starch. Whiteness improves progressively as casein is replaced by synthetic binding agents.

In testing the efficiency of FWAs for coating compositions, conditions must be standardized carefully if results are to be comparable.

If really high-grade white-coated paper is required, addition of FWA to the pulp is essential.

5. **Other Uses of FWAs**

There is a comparatively small but significant outlet for specialized FWAs in the plastics and coatings fields. The performance of FWAs depends upon the nature of the polymeric material, and selection of a suitable FWA may present problems.

Polymeric materials are often whitened in the melt. FWAs must not contain functional groups which react with the polymer. They should not interact in any way with additives. They must resist degradation by light and possess high thermal stability.

Examples of FWAs used for whitening polymeric materials were given in the section dealing with polyester fibres (see page 71).

Fluorescent whitening agents have found uses in many other fields, albeit in very small quantities. Examples include cosmetics, photographic emulsions, dentures, waxes, engineering industries (for detection of cracks), postage stamps (for mechanical sorting), legal tenders and documents and foods.

4

EVALUATION OF WHITENESS AND FLUORESCENT WHITENING AGENTS

1. Factors Affecting Evaluation of Whiteness

A body has "perfect whiteness" when it reflects all the light in the visible spectrum. The use of FWAs has familiarized the modern consumer with the concept of a whiteness that is "whiter than white", i.e. a reflectance of over 100 per cent. A neutral "perfect white", therefore, is not necessarily the most desirable form of whiteness; whites with a higher reflectance in the region 420–460 mμ are commonly preferred.

FWAs can be evaluated realistically only by visual comparison of one white substrate with another. The perception of white, like smell or taste, is a psychological as well as physiological phenomenon. The exact nature of white preferred by individuals may vary considerably. Certain physiological factors must be taken into account, including for example the fact that the yellow or brownish pigment of the human eye increases with age; in consequence, the response to the blue region of the spectrum of individuals of different age groups may be different[(126)]. A wide range of opinion must be canvassed to provide an "average observer response" in assessing a standard of whiteness that is preferred. Such a standard is valid only under the conditions used in making the assessment, and is no longer meaningful if the conditions are changed.

Under standard test conditions, findings should be reproducible. The standard conditions used for visual comparison of whites may be simulated instrumentally, and instrumental evaluation techniques have been established which agree in wide measure with visual results. Nevertheless, such techniques cannot entirely replace the visual evaluation of samples treated with FWAs.

The evaluation of whites obtained by the use of FWAs is complicated by the fact that most FWAs are adding light which is not in itself perfect white; the hues (violet, blue or green) are distinguished merely by a slight shift of the fluorescence spectrum towards the violet or green region. The human eye associates blueness with whiteness, but it is also most sensitive in the green region. Despite these complications, it is possible to control whiteness instrumentally with reference to a preferred hue of white established by "average observer response".

Individual preference for one particular white over another depends on the following physical and other factors: (*a*) light source and geometry of illumination, (*b*) eye response

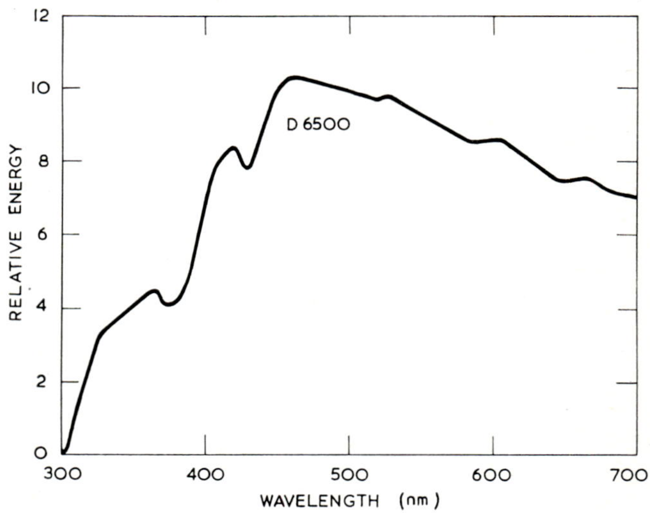

Fig. 4.1. The spectral distribution of artificial daylight source for the evaluation of FWAs

(After Judd and coworkers, 127)

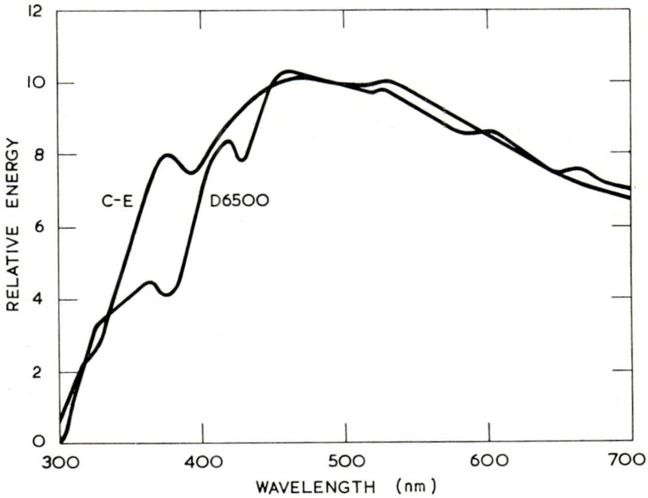

Fig. 4.2. Spectral distribution of "Color-Eye"—lamp source used in "Color-Eye" colorimeter/spectrophotometer

(After W. A. Coppock, "Pulp and Paper Magazine Canada", 1966, T-504–505

and geometry of observation, (*c*) properties of the substrates, and (*d*) physiology and psychology of hue preference.

1.1. *Light Source and Geometry of Illumination and Observation*

The most accurate methods of colour measurement assess the percentage re-emission in relation to wavelengths of the visible light. The establishment of a suitable artificial-daylight source for the evaluation of FWAs has been discussed by many authors, and spectral distribution of such a source has been defined by Judd and Coworkers[127] (Fig. 4.1).

An ideal white surface reflects diffusely in all directions. Instruments which differ in the geometry of their illumination and viewing characteristics can give different results with substrates which do not reflect uniformly in all directions.

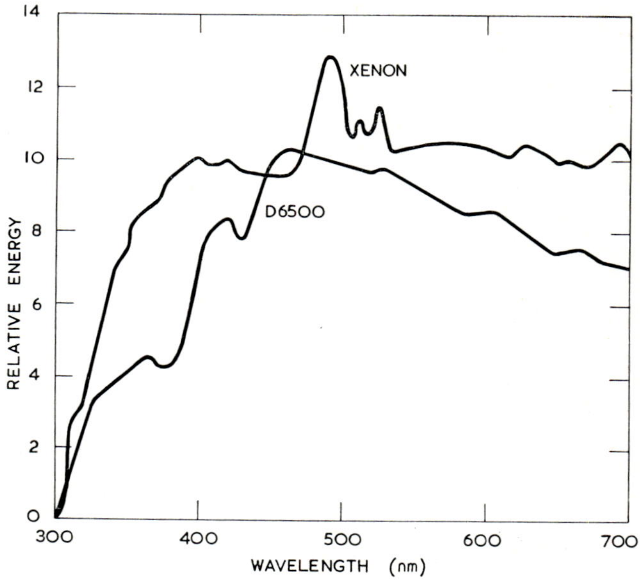

Fig. 4.3. Spectral distribution of "Xenon Lamp"—used in Zeiss Elrepho colorimeter. See Fig. 4.2

The angular conditions recommended for the colorimetry of an opaque specimen are that the light shall strike the specimen at 45° and the specimen shall be viewed along the perpendicular to its surface. These conditions avoid inclusion of the main beam of specularly-reflected light; they should be used particularly for glossy and semi-glossy material, and in cases where appearance is the chief concern.

The diffuse sphere is also widely used, however, the specimen being illuminated along or near the perpendicular, and most of the reflected light being collected, regardless of direction. This technique avoids any waste of the reflected flux; it is thus well suited for use with photoelectric devices having a small margin of sensitivity.

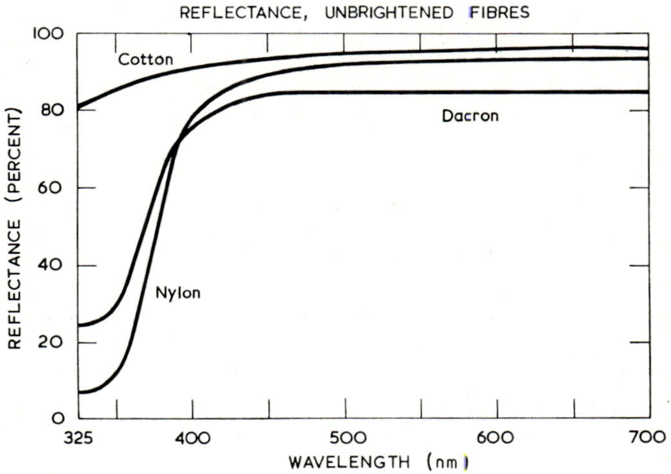

Fig. 4.4. UV and visible light reflectance curves

(After P. S. Stensby, "Soap and Chemical Specialities" May 1967, p. 90)

1.2. *Substrates*

FWAs absorb near-UV light, and substrates that absorb in this region will thus compete with an FWA by absorbing energy in this wavelength region; they diminish the effectiveness of the FWA. UV and visible reflectance curves of some fibres are given in Fig. 4.4.

If the substrate absorbs strongly in the visible region (e.g. if the substrate is yellow), the total reflectance will be low and the addition of emitted light contributed by the FWA will not be sufficient to produce a satisfactory white. To obtain a good white with the help of an FWA, it is essential to begin with a chemically-bleached ground.

Surface properties, such as gloss and transparency, are factors which must also be considered when evaluating the effect of FWAs.

1.3. *Eye Response and Preference*

Physical factors involved in the evaluation of FWAs can be standardized, but the preference of an average consumer can be determined only by carrying out a visual evaluation of a large number of whites, using a panel of many observers. An evaluation of this sort may establish an "average observer response", the value of the assessment depending upon the nature of the consumer "sample" used.

There is little precise information available regarding the hue preferred by the average observer. In North America and in many western European countries a slight violet cast apparently is preferred; in other parts of the world, consumers prefer a slightly different hue. People tend to equate whiteness with blueness plus brightness; as blueness increases, perception of whiteness increases until a point is reached where the blueness is perceived. As most household detergent FWAs are triazinyl stilbenes, the average observer is conditioned to the hue of white displayed by linen in daily use, which has a violet hue.

A number of authors have dealt with this subject, and Stensby[128] and Vaeck[129] have summarized the results as follows: "observed whiteness is a complex but perfectly rational phenomenon and that besides the colour specification of the near white surfaces, many other factors have an influence". The "other factors" include physical structure of the substrate (gloss, transparency, etc.) and personal preference.

2. **Practical Evaluation of Whiteness**

2.1. *Visual*

The most direct method of evaluating whiteness is by visual examination. Fabric samples to be compared are viewed under natural north daylight inside a window, or under artificial daylight using special lamp combinations supplied

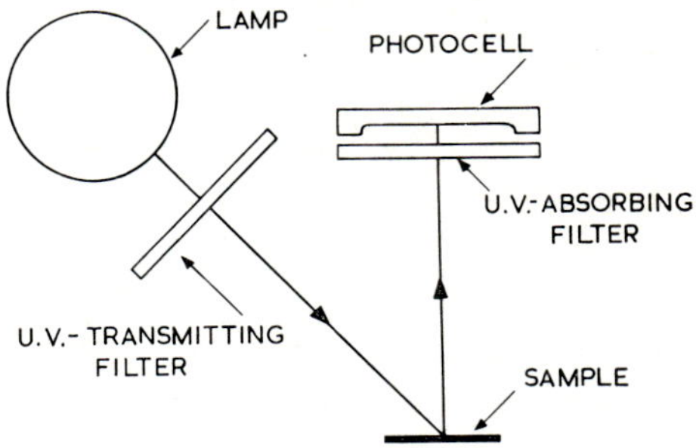

Fig. 4.5. General layout of fluorimeter used for measurement of gross fluorescence of substrates

(After J. M. Adams, "J. Soc. Dyers and Colorists", 1961, 77, 675)

by several companies, including, for example, Leslie Hubble Ltd., Leicester, U.K. (Verivide Colour matching cabinets) or Macbeth Daylighting Corp., Newburgh, N.Y., U.S.A. The colours around the sample may affect the observed results, and it is preferable to use grey as the surrounding colour. The greater the number of observers making the observation, the better. Untrained observers may be preferable to experienced observers, as the latter tend to be influenced by previous experiences. Coppock[(130)] indicates that women, who are seldom colour blind and have better self-correlation than men, probably make better observers.

The disadvantages of visual evaluation include (*a*) lack of precision, and (*b*) unreliability of results in the comparison of FWAs with different hues, because of individual hue preference. Visual examination under UV light gives results of increased precision when FWAs of the same type are being examined[(131)].

2.2. *Instrumental*

Fluorimetric Methods. Fluorimetric methods are convenient for measuring the gross fluorescence of substrates. The general layout of a typical fluorimeter is shown in Fig. 4.5.

Fluorimeters are used to determine the intensity of fluorescence of substrates treated with FWAs. They provide an accurate method of comparing strengths of fluorescence, which is useful for quality control and checks on build-up, fastness properties, etc. This method cannot be used very effectively, however, for comparing FWAs of different hues. A combination of the fluorimetric method with visual appraisal provides the experienced observer with a useful guide in the evaluation of different FWAs.

The fluorimetric method can be used, especially at low levels of application (below saturation level), to determine the amount of FWA taken up by different fabrics, by making use of the Kubelka-Munk relationship[132]:

$$C \propto \frac{(1 - R)^2}{2R} - \frac{(1 - Ro)^2}{2Ro} = \frac{K}{S}$$

where C = concentration of FWA on the fabric.
R = reflectance of the dyed fabric at 365 mμ.
Ro = reflectance of the undyed fabric at 365 mμ.
K = coefficient of absorption of the FWA at 365 mμ.
S = scattering constant of the fabric.

Spectrophotometric and Colorimetric Methods. A disadvantage of visual appraisal lies in the problem of communicating the results of an assessment in quantitative physical units. Quantitative measurements can be made, however, by using suitably-modified spectrophotometers. Tests are carried out by irradiating a sample of the fabric with light which has characteristics similar to daylight (Figs. 4.2, 4.3). The colour characteristics and luminosity of the reflected light are compared with those of light reflected by standard white (MgO).

Reflectance curves of fabric samples treated with different concentrations of FWA may be plotted; examples are shown in Fig. 4.7. If all FWAs emitted light of exactly the same spectral characteristics, it would be a relatively simple

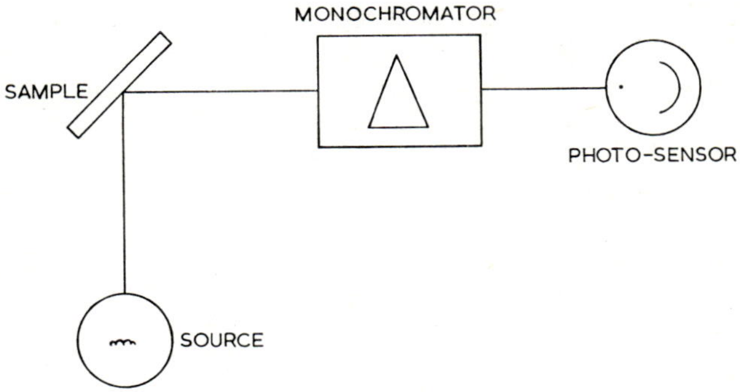

Fig. 4.6. Spectrophotometric Methods

(After W. A. Coppock, "Pulp and Paper Magazine Canada", 1966, T-505)

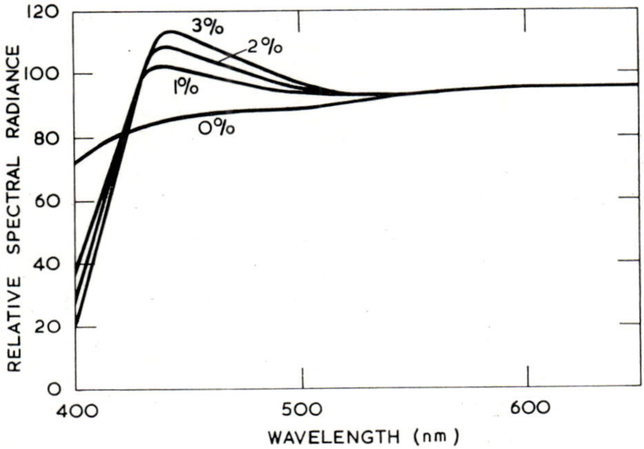

Fig. 4.7. Reflectance curves of fabric samples treated with different concentrations of FWA

(After W. A. Coppock, "Pulp and Paper Magazine Canada", 1966, T-506)

matter to correlate the reflectance measurements with visual assessment. FWAs in addition to making the substrates lighter, however, also impart a characteristic hue which varies with their chemical structure.

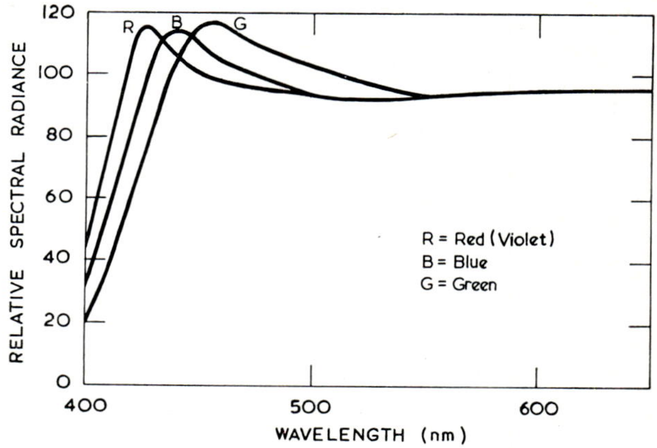

Fig. 4.8. Reflectance curves of a red, a blue and a green shade FWA

(After W. A. Coppock, "Pulp and Paper Magazine Canada," 1966, T-505)

Fig. 4.8 shows the reflectance curves of FWAs of red, blue and green shades.

Colorimeters measure in physical terms the response of an average eye towards coloured objects. C.I.E. (Commission International de l'Eclairage) carried out a detailed study of the response of the average eye to colours, and in 1931 defined the average observer response to a colour in terms of three primaries—tristimulus $\overline{X}$, $\overline{Y}$, $\overline{Z}$ values. $\overline{X}$ represents the observer's response to the red primary, $\overline{Y}$ to the green and $\overline{Z}$ to the blue. The $\overline{Y}$ also represents the observer's response to luminosity (Fig. 4.9).

Using a spectrophotometer of the type described, a reflectance spectrum is plotted. The values of reflectance (from the curve) for a series of definite wavelengths are taken and added. The sum is multiplied by a factor which directly yields $\overline{X}$, $\overline{Y}$, $\overline{Z}$ values.

Selected filters may be used with many instruments (colorimeters) (Fig. 4.10). The $\overline{X}$, $\overline{Y}$, $\overline{Z}$ values are again obtained by multiplying by predetermined factors. These

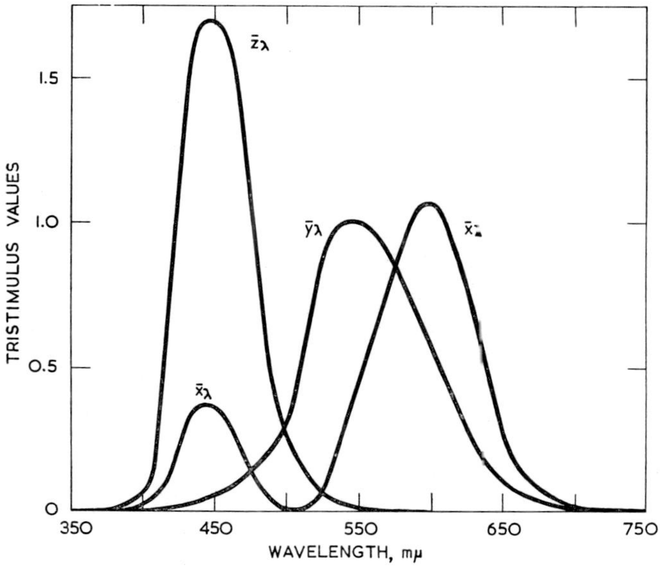

Fig. 4.9

(After Amer. Dyestuff Reporter (Nov. 21, 1966) 1966, 998)

values can be converted into one of the chromaticity systems. The $\overline{X}$, $\overline{Y}$ coordinates of the samples may be plotted in the CIE chromaticity diagram to obtain information on the hue of a white sample (Fig. 4.11).

A more detailed description of the methods employed by colourists, manufacturers and users of FWAs for the evaluation of whiteners may be obtained by reference to accounts by Stensby[(133)], Vaeck[(134)], Nieuwenhuis[(135)] and Meckel[(136)], who have summarized much of the work in this field.

Attempts have been made to correlate the instrumental value for whiteners with the results of visual assessment, and whiteness scales or functions have been suggested. Some of these are summarized in Table 4.1[(137)].

Most of the functions have limitations, and in all cases the final results must be confirmed by visual examination.

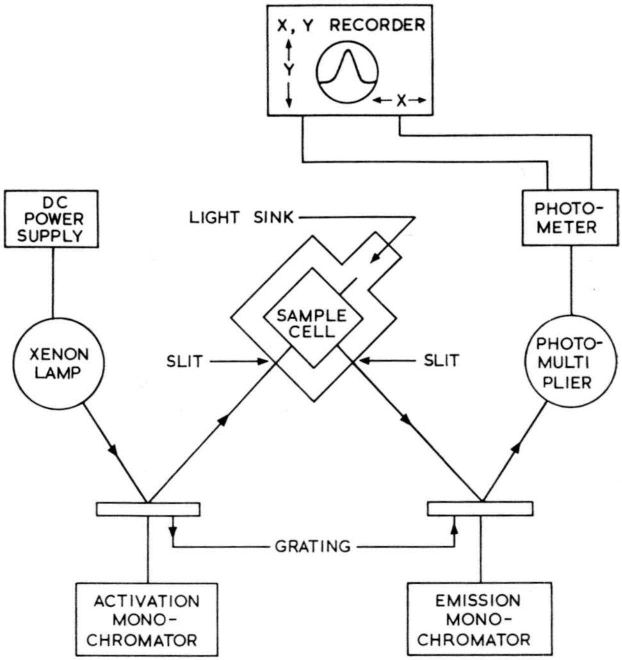

Fig. 4.10

(After W. A. Coppock loc. cit.)

However, considerable time can be saved by using such correlations internally within an organization. Stensby and Seve have suggested that attempts should now be made to derive a standardized whiteness function or scale.

Fastness Tests. Fastness tests carried out on fabrics treated with FWAs are similar to those carried out on dyed fabrics. Such tests are often of doubtful value, and attempts are being made to formulate more satisfactory techniques. Several authors have investigated the problems associated with fastness tests on fabfics treated with FWAs[138].

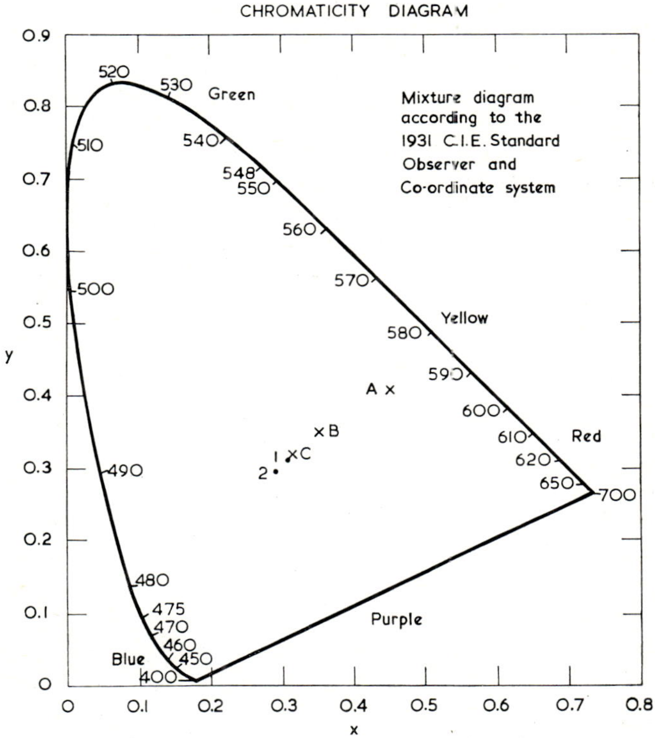

Fig. 4.11. Chromaticity Diagram. Note: A, B, and C are standard illuminants. No. 1 and No. 2 are swatches of cloth treated with FWA. No. 2 contains considerably more FWA than No. 1

(After P. S. Stensby, Soap and Chemical Specialities, May 1967, p. 130)

3. Evaluation of FWAs

Dilute solutions of FWAs absorb UV light very strongly. This property is made use of in the quantitative determination of FWAs in solution. Characteristic spectra are obtained when a UV spectrophotometer is used to measure the absorption or transmission of UV light by FWA solutions (see Fig. 4.12).

TABLE 4.1

Whiteness Functions

Functions	*Remarks*	*Authors*
1. $W = \sum_{400}^{700} P(\lambda)A(\lambda)\Delta\lambda + K$	$P(\lambda)$ = relative radiance measured at wavelength λ $A(\lambda)$ = spectrum weighting function at wavelength λ K = constant	Grum
2. $W = Y + kE$	E = chromaticity value from graph k = factor weighting luminosity against chromaticity	Vaeck
3. $W = 2B - A$	The relation between X, Y, Z	Stephansen
4. $W = G/3 + (B - A)$	and A, G, B depends on the in-	Berger
5. $W = 4B - 3G$	strument used. In one case, for	Taube
6. $W = L - 3b$	example: $X = 0{\cdot}782A + 0{\cdot}1988$;	Hunter
7. $W = L + 3a - 3b$	$Y = G$; $Z = 1{\cdot}181B$	Stensby

At low concentrations, the solutions absorb very little in the visible region; at higher concentrations, they begin to absorb violet and blue light, and thus appear yellowish. Measurements of absorption or transmission of light, expressed as optical density (D), may be used directly to determine the concentration of FWA in solution.

$D = \log(Io/I_T)$ — Io = intensity of incident light

I_T = intensity of light transmitted

$D = \text{k.Cl}$ — k = extinction coefficient

(Beer-Lambert's Law) c = concentration of FWA

l = thickness of solution through which light passes.

The optical density is measured usually at the wavelength of maximum absorption. The two values commonly quoted for quality control purposes are (*a*) k = specific extinction coefficient—i.e. optical density of unit concentration and

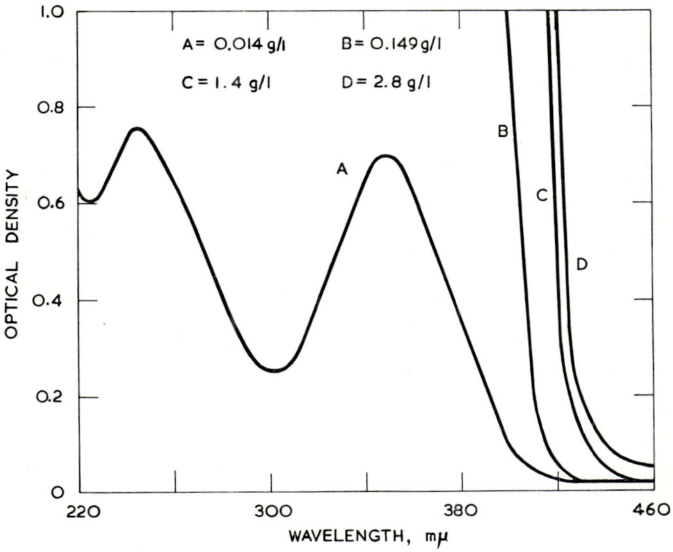

Fig. 4.12. Spectra obtained when dilute solutions of FWAs absorb UV light

thickness, and (*b*) $E_1^{1\%}$ cm, the optical density of a 1 cm layer of 1 per cent solution.

From the above relationships, it will be evident that if the specific extinction coefficient of a particular FWA is known, the determination of $E_1^{1\%}$ cm of the standardized product helps in assessing active ingredient content.

The optical density of FWAs is dependent on the solvent and the pH of the solution, and these must be standardized when comparisons are made.

Spectrofluorimeters are used to measure the emission spectra of solutions of FWAs (Fig. 4.13). The intensity at the fluorescence maximum may be used, with certain reservations, to determine the concentration of FWA in solution. At lower concentrations the UV light is incompletely absorbed; at higher concentrations, quenching occurs

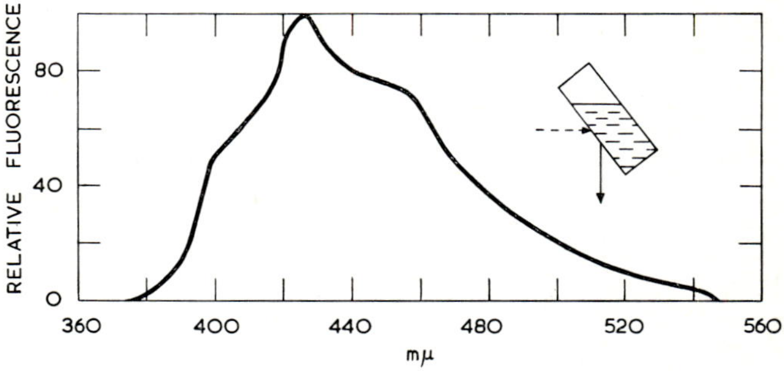

Fig. 4.13. Fluorescence emission curve of a FWA
After H. Mantz, SVF Fachorgan fur Textilveredlung, 1964, 19(6), p.478)

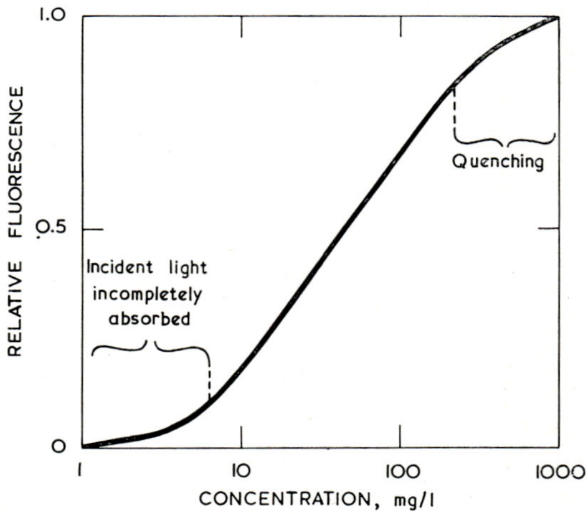

Fig. 4.14 Relation between concentration and fluorescence for solutions of a FWA[139]

(Fig. 4.14). For quantitative determination of concentration of FWAs in solution, this method compares unfavourably in respect of precision and reproducibility with the method involving measurement of absorption.

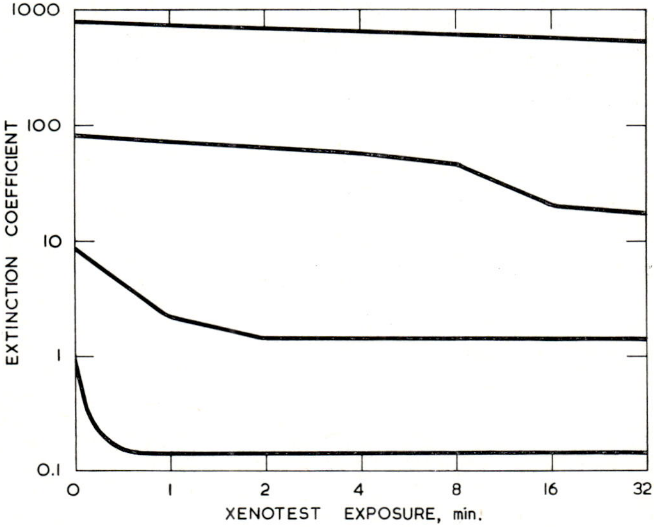

Fig. 4.15. Relation between extinction coefficient and stability to light of a triazinylamino stilbene FWA

Dilute solutions of FWAs, particularly those based on triazinylaminostilbene disulphonic acid, are very sensitive to light. When such solutions are exposed to light there is a gradual conversion of strongly-fluorescing *trans*-form to non-fluorescent *cis*-form. This phenomenon is dependent on the concentration of the solution (Fig. 4.15). At low concentrations, the fluorescence fades rapidly on exposure to light; at higher concentrations, fading is slower[139,140].

The *cis*-form reverts slowly to *trans*-form on further exposure to light. This curious phenomenon has been observed in solution, on chromatograms, and on textiles.

4. **Identification of FWAs**

The number of chemically-different FWAs appearing on the market is increasing every day. Also, many chemically-

TABLE 4.2

Eluents for Paper Chromatography of FWAs

Type of Brightener	*Eluent*	*Parts*
Coumarin	10 per cent sodium carbonate	90
	dimethylformamide	10
Oxazole	dimethylformamide	80
	formic acid (1 — N)	20
Stilbene and imidazole	*n*-butyl acetate	40
	water	15
	pyridine	40

Eluents for Thin-layer Chromatography of FWAs

Type of Brightener	*Eluent*	*Parts*
Coumarin	cyclohexane	50
	ethyl acetate	50
Oxazole	benzene	80
	chloroform	20
Stilbene	pyridine	30
	n-butanol	40
	ethanol	10
	ammonia (0·880)	20
Imidazole	carbon tetrachloride	90
	s-butyl alcohol	10

equivalent FWAs are sold in different forms and concentrations under a variety of names. Time-consuming chemical and physical analyses must be carried out to identify individual products and establish their chemical structure. Preliminary investigation of the products, however, will often indicate the presence or absence of certain structures. Tests commonly carried out include:

(*a*) spot tests,
(*b*) physical properties, e.g. solubility, etc.,

(*c*) wash tests, e.g. affinity for different fibres, hypochlorite stability, etc.,
(*d*) UV absorption and fluorescence spectra.

The FWA is often present on a substrate from which it can be extracted using a darkened Soxhlet apparatus. Chromatographic examination (paper or thin-layer) of the extract and comparison with known compounds may identify the FWA[141]. Some of the eluents used for the chromatographic analysis of FWAs are listed in Table 4.2[139].

This short monograph has dealt briefly with the chemistry, uses, analysis and other important aspects of FWAs, and the references quoted will enable the reader to study the subject in greater depth. In comparatively recent times, FWAs have become products of considerable commercial importance. The rapidly-changing patterns of progress within the textile and allied industries will necessitate continuous development of FWA chemistry and technology in the future.

BIBLIOGRAPHY

1. Anon., *Am. Dyestuff Reporter*, **57** (9), 303 (1968).
2. *Fluorescence—Theory and Practice*, Ed. G. Guilbault. Edward Arnold (1967).
3. A. v. Lagorio, *Angewandte Chemie*, **34**, 585 (1921).
4. P. Krais, *Melliand Textilberichte*, **10**, 468 (1929).
5. Ultrazell GmbH, Fr. Pat. 803,753.
6. Hoffmanns Stärkefabriken, D.R.P. 765,901.
7. Imperial Chem. Industries, B.P. 442,530, U.S.P. 2,089,413.
8. I. G. Farben, U.S.P. 2,171,427.
9. I. G. Farben, D.R.P. 752,677. D.R.P. 731,558 and Fr. Pat. 874,939.
10. *Synthetic Organic Chemicals—U.S. Production and Sales*, U.S. Tariff Commission.
11. Th. Förster, *Fluoreszenz Organischer Verbindungen*. Vandenhoeck u. Ruprecht: Gottingen, 1951, p. 98.
12. Guilbault, loc. cit., p. 45.
13. K. W. Hausser, E. Kuhn and R. Kuhn, *Zeit. Physikal. Chem. B.*, **29**, 417 (1935).
14. J. Thiele and H. Balhorn, *Ber.*, **37**, 1463 (1904).
15. H. Kauffmann, *ibid.*, **50**, 515 (1917).
16. Y. Hirschberg and F. Bergmann, *J. Amer. Chem. Soc.*, **72**, 5118 (1952).
17. A. S. Cherkasov, *Doklady Akad. Nauk S.S.S.R.*, 716, 146 (1962).
18. H. Ley and K. Engelhardt, *Zeit. Physikal. Chem.*, 1 (1910).
19. Th. Förster, loc. cit., p. 135.
20. R. Zweidler, H. Hausermann, *Kirk-Othmer's Ency. Chem. Tech.*, **3**, 737 (1964); H. W. Zusmann, *Ency. Polymer Sci. and Tech.*, **2**, 606 (1965); D. A. W. Adams, *J. Soc. Dyers and Color.*, **75**, 22 (1959); E. Uhlein, *Optischer Aufheller*. Moser Verlag, W. Germany, 1957.
21. Lever Bros., B.P. 567, 716; J. H. P. Tyman, *J. Soc. Dyers and Colorists*, 103 (1965).
22. Procter and Gamble Ltd., B.P. 663,919; Lever Bros., B.P. 647,718, B.P. 647,759; Amer. Cyanamid Co., B.P. 662,558, B.P. 700,510; DuPont, B.P. 665,415.
23. Geigy, S.A., B.P. 692,348.
24. G.A.F., U.S.P. 2,589,519.
25. Idem., U.S.P. 2,623,064.
26. DuPont, U.S.P. 2,700,046; G.A.F., B.P. 660,868.
27. DuPont, U.S.P. 2,600,004; Bayer, B.P. 760,008.
28. Geigy, S.A., B.P. 656,110.
29. *Idem.*, B.P. 1,031,750
30. F.I.A.T., F.R., 1302, The Blankophors (HMSO, London).
31. G.A.F., U.S.P. 2,720,538; U.S.P. 2,676,982.
32. S. Horrobin, *J. Chem. Soc.*, 4130 (1963); B. Bitter and H. Zollinger, *Helv. Chim. Acta*, **44**, 812 (1961).
33. Cassella Farbwerke, B.P. 814,579; Ciba, B.P. 864,900.
34. Geigy, S.A., B.P. 595,065.
35. Bayer, B.P. 894,744; Ciba, B.P. 681,642; Sandoz, Belg. P. 711,110.
36. Bayer, Belgian P. 707,448; Ciba, Belg. P. 724,507.

37. Geigy, S.A., B.P. 997,044; B.P. 1,116,619; Sterling Drug Co., B.P. 1,093,507; Sumitomo, Belg. P. 727,740.
38. I.C.I., B.P. 986,338; DuPont, B.P. 890,420; Geigy, S.A. Belg. P. 701,734.
39. I.C.I., B.P. 623,849; B.P. 624,051; Sandoz, B.P. 838,271.
40. Ciba, B.P. 696,296; I.C.I., B.P. 1,125,934.
41. Geigy, S.A., B.P. 969,402.
42. Sterling Drug Co., B.P. 997,175.
43. Geigy, S.A., B.P. 944,263.
44. Sandoz, B.P. 827,567.
45. I.C.I., B.P. 881,295, B.P. 986,338; Ciba, B.P. 882,769.
46. Ciba, Fr. P. 1,026,969; G.A.F., U.S.P. 2,713,406; I.C.I. Fr. P. 1,298,553; Nippon Chem. Ind., Jap. P. 7,079 ('57); Bayer, B.P. 1,041,530.
47. Hickson and Welch Ltd., B.P. 1,021,527; Ciba, B.P. 935,515.
48. Bayer, B.P. 715,239; B.P. 815,742; Sandoz, B.P. 760,082.
49. Geigy, S.A., B.P. 1,109,537; Farbwerke Hoechst, B.P. 1,114,570.
50. Bayer, B.P. 855,840.
51. G.A.F., U.S.P. 2,618,636; I.C.I., B.P. 645,413; B.P. 704,974; Geigy, B.P. 773,152; Sandoz, B.P. 896,533; Geigy, B.P. 973,113; B.P. 1,063,424; Sumitomo, B.P. 1,072,867.
52. Ciba, B.P. 941,993; Bayer, B.P. 841,189; Geigy, S.A., B.P. 930,393, B.P. 1,086,408, B.P. 1,129,548.
53. Amer. Cyanamid Co., Fr. Pat. 1,337,061; Ciba, B.P. 941,993.
54. Bayer, B.P. 1,010,759; G.A.F., B.P. 1,021,527.
55. Francolor, B.P. 923,122.
56. Amer. Cyanamid, U.S.P. 2,688,617.
57. Ciba, Fr. P. 1,040,777; Hickson and Welch Ltd., B.P. 796,524; Amer. Cyanamid, B.P. 721,281; Geigy, B.P. 654,028, B.P. 654,779, B.P. 692,347; Ciba, B.P. 732,139.
58. Geigy, S.A., B.P. 779,842.
59. Geigy, S.A., B.P. 666,198, B.P. 666,200.
60. B.A.S.F., B.P. 913,735, B.P. 924,762, B.P. 948,267, B.P. 1,045,443, B.P. 916,610; Pilot Chemicals, B.P. 1,043,501.
61. Ciba, Belgian P. 721,754; I.C.I. B.P. 902,022.
62. Bayer, B.P. 683,895.
63. Clayton Aniline, B.P. 716,338; Geigy, S.A., B.P. 861,451; Sandoz, B.P. 796,211; Geigy, S.A., B.P. 718,692, B.P. 779,505, B.P. 779,958; Bayer, B.P. 1,163,890.
64. Hickson and Welch Ltd., B.P. 835,779.
65. Ciba, B.P. 821,843.
66. Geigy, S.A., B.P. 717,889, B.P. 726,129, B.P. 779,505, B.P. 763,696, B.P. 1,108,865, B.P. 1,150,408; Hickson and Welch Ltd., B.P. 828,329, B.P. 855,784, B.P. 874,611; G.A.F., B.P. 1,141,921.
67. Geigy, S.A., B.P. 668,199; B.P. 869,372, Belg. P. 622,276; G.A.F., U.S.P. 2,821,550.
68. Geigy, S.A., B.P. 917,342; G.A.F., B.P. 917,880, B.P. 928,160, B.P. 932,816, B.P. 944,265, B.P. 987,456; Hickson and Welch Ltd., B.P. 875,274, B.P. 991,931, B.P. 991,932.
69. DuPont, U.S.P. 2,700,043, U.S.P. 2,700,044, U.S.P. 2,713,054, U.S.P. 2,715,632, U.S.P. 2,713,056.

70. Mitsui Chem. Ind. Ltd., U.S.P. 3,401,048.
71. H. Gold, S. Petersen and O. Bayer, in *Recent Progress in the Chemistry of Natural and Synthetic Coloring Matters and Related Fields*, Ed. T. S. Gore *et al.* Academic Press, 1962, p. 610.
72. Bayer, B.P. 829,788, B.P. 1,108,416, B.P. 1,132,314, B.P. 1,172,328, B.P. 1,172,657.
73. Bayer, B.P. 787,429, Belgian P. 713,056.
74. Hickson & Welch Ltd., B.P. 1,052,179, B.P. 1,101,156.
75. Ciba, B.P. 600,696, B.P. 611,510, B.P. 623,428, B.P. 653,681.
76. Amer. Cyanamid, U.S.P. 2,992,218, U.S.P. 3,019,221; Sterling Drug Co., U.S.P. 2,838,504, U.S.P. 2,878,248; Ciba, B.P. 853,129, B.P. 888,209, B.P. 897,095, B.P. 983,609, B.P. 990,406, Belgian P. 623,629.
77. Ciba, B.P. 941,048; Vereingte Glanzstoff, B.P. 865,804; Bayer, B.P. 1,059,256, B.P. 1,125,535.
78. Ciba, B.P. 902,058, B.P. 902,059, B.P. 835,894, B.P. 828,545, B.P. 920,753, B.P. 907,819.
79. Ciba, B.P. 1,102,931, B.P. 1,077,982, B.P. 947,981, B.P. 947,982, B.P. 947,983, B.P. 940,770, B.P. 1,063,797; Nippon KKK, B.P. 1,065,516; Sumitomo, B.P. 1,062,257; Hoechst, B.P. 1,059,687; Kodak, B.P. 1,072,914, B.P. 1,072,915, B.P. 1,072,917, B.P. 1,072,918, B.P. 1,072,919; Ciba, B.P. 981,822, B.P. 950,890.
80. Hoechst, B.P. 1,055,183, B.P. 1,055,184, B.P. 1,071,346, B.P. 1,079,236, B.P. 1,116,635.
81. Mitsui, B.P. 967,483; Amer. Cyanamid, B.P. 1,037,378.
82. Ciba, B.P. 990,397, B.P. 1,045,818, B.P. 1,113,614; Hoechst, B.P. 1,115,597; Bayer, B.P. 1,059,256; Hoechst, B.P. 1,116,635; Ciba, B.P. 1,142,499, B.P. 941,048; Amer. Cyanamid, B.P. 1,007,017; Kodak, B.P. 1,072,916; Hoechst, B.P. 1,156,800; Ciba, B.P. 1,008,456, B.P. 1,073,669, B.P. 1,115,274, B.P. 1,143,505; Daito Chem. Ind., Japan. P. 20,970 ('68).
83. Daito Chem. Ind., B.P. 1,019,575, B.P. 967,360; Ciba, B.P. 981,822; Sumitomo, Japan. P. 06,981 ('69).
84. Ciba, Belgian P. 665,400, Belgian P. 669,470, Belgian P. 680,635, B.P. 1,164,606, B.P. 1,167,092.
85. R. Anliker, H. Hefti and H. Kasperl, *J. Amer. Oil Chemists' Soc.*, **46,** 78 (1969).
86. Ciba, B.P. 899,842.
87. L. B. Holliday, B.P. 755,117; Sandoz. B.P. 741,959; Ciba, B.P. 655,258.
88. Bayer, Belgian P. 660,602; Geigy, B.P. 786,234.
89. Bayer, B.P. 933,042; Hickson & Welch Ltd., B.P. 1,152,875, B.P. 1,154,272.
90. Geigy, S.A., B.P. 885,847, B.P. 1,114,327, B.P. 991,202, B.P. 1,102,317; Bayer, Belgian P. 706,639; Geigy, S.A., Belgian P. 710,870; Bayer, B.P. 1,113,918, B.P. 1,100,579; Sandoz, B.P. 990,102; Bayer, Belgian P. 727,389, Belgian P. 647,377, Belgian P. 731,417.
91. Ciba, B.P. 914,719; Bayer, B.P. 1,010,562, Belgian P. 715,569.
92. Bayer, B.P. 1,103,531, B.P. 1,146,798.
93. Ilford Ltd., B.P. 669,590, B.P. 669,591, B.P. 670,857, B.P. 698,253, B.P. 712,764.

94. Bayer, B.P. 808,113, B.P. 832,239, B.P. 906,960; Hickson & Welch Ltd., B.P. 993,055, B.P. 1,130,882; Hoechst, B.P. 1,111,986, B.P. 1,047,226, Belgian P. 661,139; Geigy, S.A., B.P. 986,059, B.P. 1,146,487; Unilever, B.P. 883,826.
95. S. Petersen, C. W. Schellhammer and A. Wagner, *Angew. Chem.* (*Inter.*), **5** (8), 699 (1966); B. H. Chase and J. M. Evans, *J. Chem. Soc.*, 4825 (1964).
96. B.A.S.F., B.P. 741,798.
97. Bayer, B.P. 1,067,786, B.P. 1,154,995.
98. T. Kasai, B.P. 1,003,083, B.P. 1,003,084; Mitsubishi Chem. Ind., B.P. 962,019; Ugine Kuhlmann, Belgian P. 718,203; Research Inst. for Production Dev., Japan. P. 20,971 ('68).
99. Mitsubishi Chem. Ind., B.P. 1,054,436.
100. Amer. Cyanamid, U.S.P. 2,620,343, U.S.P. 2,563,795, U.S.P. 2,573,652, U.S.P. 3,226,247.
101. Bayer, Belgian P. 712,484.
102. Ilford, B.P. 669,086; Geigy, S.A., U.S.P. 2,809,123, Belgian P. 710,261; Geigy, S.A., B.P. 736,452, B.P. 1,044,436, B.P. 1,085,441.
103. I.C.I., B.P. 985,484; Unilever, B.P. 1,141,454.
104. Ciba, B.P. 1,110,806.
105. Otto Dann, B.P. 871,351.
106. I.C.I., B.P. 838,994, B.P. 851,792.
107. Geigy, S.A., B.P. 1,031,548.
108. Hoechst, B.P. 1,014,651.
109. L. B. Holliday Ltd., B.P. 722,791.
110. Geigy, S.A., B.P. 1,010,749.
111. Bayer, B.P. 846,148.
112. Sterling Drug Co., B.P. 876,694.
113. I.C.I., B.P. 1,086,068.
114. Bayer, B.P. 616,472.
115. Anon., *Soap and Chemical Specialties*, Oct. 1968, pp. 37–41.
116. P. S. Stensby, *Detergent Age*, **3** (9), 20 (1967).
117. H. Hausermann and P. Keller, *Textil Rundschau*, **16**, 178 (1961).
118. P. Carter and J. D. Dance, *Soap and Chemical Specialties*, Feb., 1969, p. 56.
119. P. S. Stensby, *Detergent Age*, Jan. and Feb., 1968, p. 24 and p. 26.
120. S. Goldwasser, *Soap and Chemical Specialties*, 1962 (Aug. and Sept.), p. 38 and p. 47.
121. W. A. Coppock, *Pulp and Paper Mag. Can.*, **67**, 499 (1964).
122. H. Mantz and A. Brüggemann, *SVF Fachorgan für Textilveredlung*, **19** (6), 426 (1964).
123. D. R. Graham and K. W. Statham, *J. Soc. Dyers and Colourists*, **72,** 434 (1956).
124. *Treatment of Woven and Knitted Wool Fabrics with Thioureaformaldehyde to Retard Sunlight Yellowing*, C.S.I.R.O. Australian Report No. G.16, March, 1968. I. H. Leaver and G. C. Ramsay, *Text. Res. J.*, **39** (8) 722–33 (1969).
125. Anon., *Amer. Dyestuff Reporter*, May 5, 1969, p. 11.
126. *Colorimetry*, National Bureau of Standards (Washington D.C.), Circular No. 478 (1950), p. 2.

127. D. B. Judd, D. L. Macadam and G. J. Wyszecki, *Opt. Soc. Am.*, 54 (8) (1964).
128. P. S. Stensby, *Soap and Chemical Specialties*, May 1967, p. 132.
129. S. V. Vaeck *et al.*, *Wäscherie-technik u. Chemie*, **17,** 26 and 898 (1964); *ibid.*, **18,** 155 (1965).
130. W. Coppock, *Amer. Dyestuff Reporter*, 1965, p. 343.
131. E. Allen, *Amer. Dyestuff Reporter*, **46,** 425 (1957).
132. *Idem.*, *J. Opt. Soc. Amer.*, **54** (4), 506 (1964).
133. P. S. Stensby, *Soap and Chemical Specialties*, May 1967, p. 84, and July 1967, p. 80.
134. S. V. Vaeck, *Annales Textiles Belg*. 96, (1966) (1–3)
135. K. J. Nieuwenhuis, *J. Amer. Oil Chemists Soc.*, **45,** 37 (1968)
136. L. Meckel, H. Milster and H. Terstiege, *Melliand Textilberichte*, 1969, p. 859.
137. P. S. Stensby, *Soap and Chemical Specialties*, July 1967, p. 86.
138. H. Heuberger, *Textilveredlung*, **4** (2), 101 (1969); G. Anders, *ibid.*, **3** (3), 116 (1968); A. Wagner and F. Gund, *ibid.*, **1** (3), 106 (1966); A. Räuchle and W. Schramm, *ibid.*, **2** (9), 719 (1967).
139. J. Lanter, *J. Soc. Dyers and Colourists*, **82,** 125 (1966).
140. H. Theidel, *Melliand Textilberichte*, 1964, p. 514.
141. V. J. Kurz and M. Schuierer, *Fette, Seifen Anstrichmittel*, **69,** 24 (1967).

APPENDIX

The following list includes some of the leading manufacturers of FWAs, and gives the trade names under which their products are sold:

	Name of Company	*Trade Names of FWAS*
1.	American Cyanamid Co., U.S.A.	Calcofluor White
2.	Badische Anilin U. Soda Fabric, (BASF), W. Germany	Ultraphor and Palanil Brilliant White
3.	C.F.M.C. (Francolor), France	Fluotex
4.	Ciba, S.A., Switzerland	Uvitex
5.	Daito Chem. Industries, Japan	Daitophor
6.	Delfland Produkten N.V., Holland	Delft White
7.	DuPont de Nemours, U.S.A.	Pontamine White
8.	Farbenfabriken Bayer, A.G., W. Germany	Blankophor
9.	Farbwerke Hoechst, A.G., W. Germany	Hostalux
10.	GAF Corp., U.S.A.	Blancophor and Tintofen
11.	J. R. Geigy, S.A., Switzerland	Tinopal
12.	Hickson & Welch Ltd., U.K.	Photine
13.	Hilton Davis Chem. Co., U.S.A.	Hiltamine Arctic White
14.	Imperial Chemical Industries Ltd., U.K.	Fluolite
15.	Melle-Bezons, France	Celumyl
16.	Mitsubishi Chem. Ind., Japan	Mikawhite
17.	Mitsui Chem. Ind., Japan	Mikephor
18.	Nippon Kayaku Co., Japan	Kayaphor
19.	Sandoz, A.G., Switzerland	Leucophor
20.	Showa Chem. Ind., Japan	Hakkol
21.	SIGMA, Italy	Optiblanc
22.	Sumitomo Chem. Co. Ltd., Japan	Whitex
23.	Verona Chem. Corp., U.S.A.	Phorwhite
24.	Viochrome Ltd., Greece	Viophos

INDEX